Robert S. Ball

Experimental Mechanics

A course of lectures delivered at the Royal College of Science for Ireland

Robert S. Ball

Experimental Mechanics
A course of lectures delivered at the Royal College of Science for Ireland

ISBN/EAN: 9783337034344

Printed in Europe, USA, Canada, Australia, Japan

Cover: Foto ©berggeist007 / pixelio.de

More available books at **www.hansebooks.com**

EXPERIMENTAL MECHANICS

EXPERIMENTAL MECHANICS

A COURSE OF LECTURES

DELIVERED AT THE ROYAL COLLEGE OF SCIENCE FOR IRELAND

BY

SIR ROBERT STAWELL BALL, LL.D., F.R.S.

ASTRONOMER ROYAL OF IRELAND

FORMERLY PROFESSOR OF APPLIED MATHEMATICS AND MECHANISM IN THE ROYAL COLLEGE OF SCIENCE FOR IRELAND (SCIENCE AND ART DEPARTMENT)

WITH ILLUSTRATIONS

SECOND EDITION

London

MACMILLAN AND CO.

AND NEW YORK

1888.

The Right of Translation and Reproduction is reserved

RICHARD CLAY AND SONS, LIMITED,
LONDON AND BUNGAY.

The First Edition was printed in 1871.

PREFACE.

I HERE present the revised edition of a course of lectures on Experimental Mechanics which I delivered in the Royal College of Science at Dublin eighteen years ago. The audience was a large evening class consisting chiefly of artisans.

The teacher of Elementary Mechanics, whether he be in a Board School, a Technical School, a Public School, a Science College, or a University, frequently desires to enforce his lessons by exhibiting working apparatus to his pupils, and by making careful measurements in their presence.

He wants for this purpose apparatus of substantial proportions visible from every part of his lecture room. He wants to have it of such a universal character that he can produce from it day after day combinations of an ever-varying type. He wishes it to be composed of well-designed and well-made parts that shall be strong and durable, and that will not easily get out of order. He wishes those parts to be such that even persons not specially trained in manual skill shall presently learn how to combine them with good effect. Lastly, he desires to economize his money in the matters of varnish, mahogany, and glass cases.

I found that I was able to satisfy all these requirements by a suitable adaptation of the very ingenious system of mechanical apparatus devised by the late Professor Willis of Cambridge. The elements of the system I have briefly described in an Appendix, and what adaptations I have made of it are shown in almost every page and every figure of the book.

In revising the present edition I have been aided by my friends Mr. G. L. Cathcart, the Rev. M. H. Close, and Mr. E. P. Culverwell.

ROBERT S. BALL.

OBSERVATORY, CO. DUBLIN,
3rd August, 1888.

TABLE OF CONTENTS.

LECTURE I.
THE COMPOSITION OF FORCES.

PAGE

Introduction.—The Definition of Force.—The Measurement of Force.—Equilibrium of Two Forces.—Equilibrium of Three Forces.—A Small Force can sometimes balance Two Larger Forces 1

LECTURE II.
THE RESOLUTION OF FORCES.

Introduction.—One Force resolved into Two Forces.—Experimental Illustrations.—Sailing.—One Force resolved into Three Forces not in the same Plane.—The Jib and Tie-rod. . . . 16

LECTURE III.
PARALLEL FORCES.

Introduction.—Pressure of a Loaded Beam on its Supports.—Equilibrium of a Bar supported on a Knife-edge.—The Composition of Parallel Forces.—Parallel Forces acting in opposite directions.—The Couple.—The Weighing Scales 34

LECTURE IV.
THE FORCE OF GRAVITY.

Introduction.—Specific Gravity.—The Plummet and Spirit Level.—The Centre of Gravity.—Stable and Unstable Equilibrium.—Property of the Centre of Gravity in a Revolving Wheel 50

TABLE OF CONTENTS.

LECTURE V.

THE FORCE OF FRICTION.

The Nature of Friction.—The Mode of Experimenting.—Friction is proportional to the pressure.—A more accurate form of the Law.—The Coefficient varies with the weights used.—The Angle of Friction.—Another Law of Friction.—Concluding Remarks 65

LECTURE VI.

THE PULLEY.

Introduction.—Friction between a Rope and an Iron Bar.—The Use of the Pulley.—Large and Small Pulleys.—The Law of Friction in the Pulley.—Wheels.—Energy 85

LECTURE VII.

THE PULLEY-BLOCK.

Introduction.—The Single Movable Pulley.—The Three-sheave Pulley-block.—The Differential Pulley-block.—The Epicycloidal Pulley-block 99

LECTURE VIII.

THE LEVER.

The Lever of the First Order.—The Lever of the Second Order.—The Shears.—The Lever of the Third Order 119

LECTURE IX.

THE INCLINED PLANE AND THE SCREW.

The Inclined Plane without Friction.—The Inclined Plane with Friction.—The Screw.—The Screw-jack.—The Bolt and Nut 131

TABLE OF CONTENTS.

LECTURE X.

THE WHEEL AND AXLE.

PAGE

Introduction.—Experiments upon the Wheel and Axle.—Friction upon the Axle.—The Wheel and Barrel.—The Wheel and Pinion.—The Crane.—Conclusion 149

LECTURE XI.

THE MECHANICAL PROPERTIES OF TIMBER.

Introduction.—The General Properties of Timber.—Resistance to Extension.—Resistance to Compression.—Condition of a Beam strained by a Transverse Force 169

LECTURE XII.

THE STRENGTH OF A BEAM.

A Beam free at the Ends and loaded in the Middle.—A Beam uniformly loaded.—A Beam loaded in the Middle, whose Ends are secured.—A Beam supported at one end and loaded at the other 188

LECTURE XIII.

THE PRINCIPLES OF FRAMEWORK.

Introduction.—Weight sustained by Tie and Strut.—Bridge with Two Struts.—Bridge with Four Struts.—Bridge with Two Ties.—Simple Form of Trussed Bridge 203

LECTURE XIV.

THE MECHANICS OF A BRIDGE.

Introduction.—The Girder.—The Tubular Bridge.—The Suspension Bridge 218

LECTURE XV.

THE MOTION OF A FALLING BODY.

Introduction.—The First Law of Motion.—The Experiment of Galileo from the Tower of Pisa.—The Space is proportional to the Square of the Time.—A Body falls 16' in the First Second.—The Action of Gravity is independent of the Motion of the Body.—How the Force of Gravity is defined.—The Path of a Projectile is a Parabola 230

LECTURE XVI.

INERTIA.

Inertia.—The Hammer.—The Storing of Energy.—The Flywheel.—The Punching Machine 250

LECTURE XVII.

CIRCULAR MOTION.

The Nature of Circular Motion.—Circular motion in Liquids.—The Applications of Circular Motion.—The Permanent Axes 267

LECTURE XVIII.

THE SIMPLE PENDULUM.

Introduction.—The Circular Pendulum.—Law connecting the Time of Vibration with the Length.—The Force of Gravity determined by the Pendulum.—The Cycloid 284

LECTURE XIX.

THE COMPOUND PENDULUM AND THE COMPOSITION OF VIBRATIONS.

The Compound Pendulum.—The Centre of Oscillation.—The Centre of Percussion.—The Conical Pendulum.—The Composition of Vibrations 299

TABLE OF CONTENTS. xiii

LECTURE XX.

THE MECHANICAL PRINCIPLES OF A CLOCK.
PAGE

Introduction.—The Compensating Pendulum.—The Escapement.
—The Train of Wheels.—The Hands.—The Striking Parts . 318

APPENDIX I.

The Method of Graphical Construction 339
The Method of Least Squares 342

APPENDIX II.

Details of the Willis Apparatus used in illustrating the foregoing
lectures 345

INDEX 355

EXPERIMENTAL MECHANICS

EXPERIMENTAL MECHANICS.

LECTURE I.

THE COMPOSITION OF FORCES.

Introduction.—The Definition of Force.—The Measurement of Force.—Equilibrium of Two Forces.—Equilibrium of Three Forces.—A Small Force can sometimes balance Two Larger Forces.

INTRODUCTION.

1. I SHALL endeavour in this course of lectures to illustrate the elementary laws of mechanics by means of experiments. In order to understand the subject treated in this manner, you need not possess any mathematical knowledge beyond an acquaintance with the rudiments of algebra and with a few geometrical terms and principles. But even to those who, having an acquaintance with mathematics, have by its means acquired a knowledge of mechanics, experimental illustrations may still be useful. By actually seeing the truth of results with which you are theoretically familiar, clearer conceptions may be produced, and perhaps new lines of thought opened up. Besides, many of the mechanical principles which lie rather beyond the scope of elementary works on the subject are very susceptible of

being treated experimentally; and to the consideration of these some of the lectures of this course will be devoted.

Many of our illustrations will be designedly drawn from very commonplace sources: by this means I would try to impress upon you that mechanics is not a science that exists in books merely, but that it is a study of those principles which are constantly in action about us. Our own bodies, our houses, our vehicles, all the implements and tools which are in daily use—in fact all objects, natural and artificial, contain illustrations of mechanical principles. You should acquire the habit of carefully studying the various mechanical contrivances which may chance to come before your notice. Examine the action of a crane raising weights, of a canal boat descending through a lock. Notice the way a roof is made, or how it is that a bridge can sustain its load. Even a well-constructed farm-gate, with its posts and hinges, will give you admirable illustrations of the mechanical principles of frame-work. Take some opportunity of examining the parts of a clock, of a sewing-machine, and of a lock and key; visit a saw-mill, and ascertain the action of all the machines you see there; try to familiarize yourself with the principles of the tools which are to be found in any workshop. A vast deal of interesting and useful knowledge is to be acquired in this way.

THE DEFINITION OF FORCE.

2. It is necessary to know the answer to this question, What is a force? People who have not studied mechanics occasionally reply, A push is a force, a steam-engine is a force, a horse pulling a cart is a force, gravitation is a force, a movement is a force, &c., &c. The true definition of force is *that which tends to produce or to destroy motion.* You

THE DEFINITION OF FORCE.

may probably not fully understand this until some further explanations and illustrations shall have been given ; but, at all events, put any other notion of force out of your mind. Whenever I use the word Force, do you think of the words " something which tends to produce or to destroy motion," and I trust before the close of the lecture you will understand how admirably the definition conveys what force really is.

3. When a string is attached to this small weight, I can, by pulling the string, move the weight along the table. In this case, there is something transmitted from my hand along the string to the weight in consequence of which the weight moves: that something is a force. I can also move the weight by pushing it with a stick, because force is transmitted along the stick, and makes itself known by producing motion. The archer who has bent his bow and holds the arrow between his finger and thumb feels the string pulling until the impatient arrow darts off. Here motion has been produced by the force of elasticity in the bent bow. Before he released the arrow there was no motion, yet still the bow was exerting force and *tending* to produce motion. Hence in defining force we must say " that which *tends* to produce motion," whether motion shall actually result or not.

4. But forces may also be recognized by their capability or tendency to prevent or to destroy motion. Before I release the arrow I am conscious of exerting a force upon it in order to counteract the pull of the string. Here my force is merely manifested by *destroying the motion* that, if it were absent, the bow would produce. So when I hold a weight in my hand, the force exerted by my hand destroys the motion that the weight would acquire were I to let it fall; and if a weight greater than I could support were placed in my hand, my efforts to sustain it would still be

properly called force, because they *tended* to destroy motion, though unsuccessfully. We see by these simple cases that a force may be recognized either by producing motion or by trying to produce it, by destroying motion or by tending to destroy it; and hence the propriety of the definition of force must be admitted.

THE MEASUREMENT OF FORCE.

5. As forces differ in magnitude, it becomes necessary to establish some convenient means of expressing their measurements. The pressure exerted by one pound weight at London is the standard with which we shall compare other forces. The piece of iron or other substance which is attracted to the earth with this force in London, is attracted to the earth with a greater force at the pole and a less force at the equator; hence, in order to define the standard force, we have to mention the locality in which the pressure of the weight is exerted.

It is easy to conceive how the magnitude of a pushing or a pulling force may be described as equivalent to so many pounds. The force which the muscles of a man's arm can exert is measured by the weight which he can lift. If a weight be suspended from an india-rubber spring, it is evident the spring will stretch so that the weight pulls the spring and the spring pulls the weight; hence the number of pounds in the weight is the measure of the force the spring is exerting. In every case the magnitude of a force can be described by the number of pounds expressing the weight to which it is equivalent. There is another but much more difficult mode of measuring force occasionally used in the higher branches of mechanics (Art. 497), but the simpler method is preferable for our present purpose.

EQUILIBRIUM OF TWO FORCES.

6. The straight line in which a force tends to move the body to which it is applied is called the direction of the force. Let us suppose, for example, that a force of 3 lbs. is applied at the point A, Fig. 1, tending to make A move in the direction AB. A standard line C of certain length is to be taken. It is supposed that a line of this length represents a force of 1 lb.

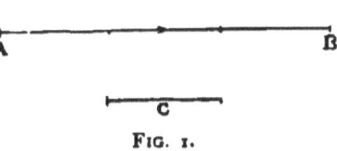

FIG. 1.

The line AB is to be measured, equal to three times C in length, and an arrow-head is to be placed upon it to show the direction in which the force acts. Hence, by means of a line of certain length and direction, and having an arrow-head attached, we are able completely to represent a force.

EQUILIBRIUM OF TWO FORCES.

7. In Fig. 2 we have represented two equal weights to which strings are attached; these strings, after passing over pulleys, are fastened by a knot C. The knot is pulled by equal and opposite forces. I mark off parts CD, CE, to indicate the forces; and since there is no reason why C should move to one side more than the other, it remains at rest.

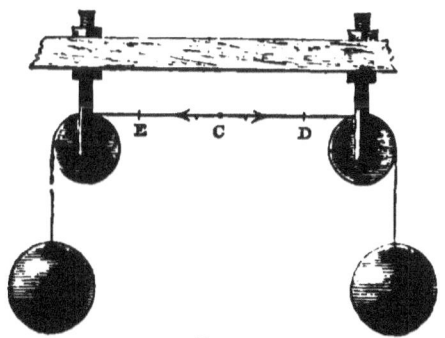

FIG. 2.

Hence, we learn that two equal and directly opposed forces counteract each other, and each may be regarded as destroying the motion which the other is striving to produce. If I make the weights unequal by adding to one of them, the knot is no longer at rest; it instantly begins to move in the direction of the larger force.

8. When two equal and opposite forces act at a point, they are said to be in *equilibrium*. More generally this word is used with reference to any set of forces which counteract each other. When a force acts upon a body, at least one more force must be present in order that the body should remain at rest. If two forces acting on a point be not opposite, they will not be in equilibrium; this is easily shown by pulling the knot c in Fig. 2 downwards. When released, it flies back again. This proves that if two forces be in equilibrium their directions must be opposite, for otherwise they will produce motion. We have already seen that the two forces must be equal.

A book lying on the table is at rest. This book is acted upon by two forces which, being equal and opposite, destroy each other. One of these forces is the gravitation of the earth, which tends to draw the book downwards, and which would, in fact, make the book fall if it were not sustained by an opposite force. The pressure of the book on the table is often called the *action*, while the resistance offered by the table is the force of *reaction*. We here see an illustration of an important principle in nature, which says that *action and reaction are equal and opposite*.

EQUILIBRIUM OF THREE FORCES.

9. We now come to the important case where three forces act on a point: this is to be studied by the apparatus represented in Fig. 3. It consists essentially of two pulleys

I.] EQUILIBRIUM OF THREE FORCES. 7

H, H, each about 2″ diameter,[1] which are capable of turning very freely on their axles; the distance between these pulleys is about 5′, and they are supported at a height of 6′

Fig. 3.

[1] We shall often, in these lectures, represent feet or inches in the manner usual among practical men—1′ is one foot, 1″ is one inch. Thus, for example, 3′ 4″ is to be read "three feet four inches." When it is necessary to use fractions we shall always employ decimals. For example, 0″·5 is the mode of expressing a length of half an inch; 3′ 1″·9 is to be read "three feet one inch and nine-tenths of an inch."

by a frame, which will easily be understood from the figure. Over these pulleys passes a fine cord, 9′ or 10′ long, having a light hook at each of the ends E,F. To the centre of this cord D a short piece is attached, which at its free end G is also furnished with a hook. A number of iron weights, 0·5 lb., 1 lb., 2 lbs., &c., with rings at the top, are used; one or more of these can easily be suspended from the hooks as occasion may require.

10. We commence by placing one pound on each of the hooks. The cords are first seen to make a few oscillations and then to settle into a definite position. If we disturb the cords and try to move them into some new position they will not remain there; when released they will return to the places they originally occupied. We now concentrate our attention on the central point D, at which the three forces act. Let this be represented by O in Fig. 4, and the lines OP, OQ, and OS will be the directions of the three cords.

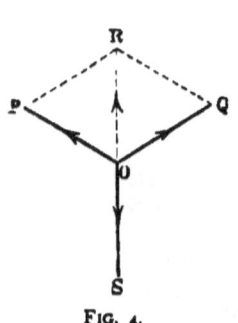

FIG. 4.

On examining these postions, we find that the three angles P O S, Q O S, P O Q, are all equal. This may very easily be proved by holding behind the cords a piece of cardboard on which three lines meeting at a point and making equal angles have been drawn; it will then be seen that the cords coincide with the three lines on the cardboard.

11. A little reflection would have led us to anticipate this result. For the three cords being each stretched by a tension of a pound, it is obvious that the three forces pulling at O are all equal. As O is at rest, it seems obvious that the three forces must make the angles equal, for suppose that one of the angles, P O Q for instance, was less than either of

I.]   EQUILIBRIUM OF THREE FORCES.   9

the others, experiment shows that the forces O P and O Q would be too strong to be counteracted by O S. The three angles must therefore be equal, and then the forces are arranged symmetrically.

12. The forces being each 1 lb., mark off along the three lines in Fig. 4 (which represent their directions) three equal parts O P, O Q, O S, and place the arrowheads to show the direction in which each force is acting; the forces are then completely represented both in position and in magnitude.

Since these forces make equilibrium, each of them may be considered to be counteracted by the other two. For example, O S is annulled by O Q and O P. But O S could be balanced by a force O R equal and opposite to it. Hence O R is capable of producing by itself the same effect as the forces O P and O Q taken together. Therefore O R is equivalent to O P and O Q. Here we learn the important truth that two forces not in the same direction can be replaced by a single force. The process is called the *composition of forces*, and the single force is called the *resultant* of the two forces. O R is only one pound, yet it is equivalent to the forces O P and O Q together, each of which is also one pound. This is because the forces O P and O Q partly counteract each other.

13. Draw the lines P R and Q R; then the angles P O R and Q O R are equal, because they are the supplements of the equal angles P O S and Q O S; and since the angles P O R and Q O R together make up one-third of four right angles, it follows that each of them is two-thirds of one right angle, and therefore equal to the angle of an equilateral triangle. Also O P being equal to O Q and O R common, the triangles O P R and O Q R must be equilateral. Therefore the angle P R O is equal to the angle R O Q; thus P R is parallel to O Q: similarly Q R is parallel to O P; that is, O P R Q is a parallelo-

gram. Here we first perceive the great law that the resultant of two forces acting at a point is the diagonal of a parallelogram, of which they are the two sides.

14. This remarkable geometrical figure is called the *parallelogram of forces*. Stated in its general form, the property we have discovered asserts that two forces acting at a point have a resultant, and that this resultant is represented both in magnitude and in direction by the diagonal of the parallelogram, of which two adjacent sides are the lines which represent the forces.

15. The parallelogram of forces may be illustrated in various ways by means of the apparatus of Fig. 3. Attach, for example, to the middle hook G 1·5 lb., and place 1 lb. on each of the remaining hooks E, F. Here the three weights are not equal, and symmetry will not enable us, as it did in the previous case, to foresee the condition which the cords will assume; but they will be observed to settle in a definite position, to which they will invariably return if withdrawn from it.

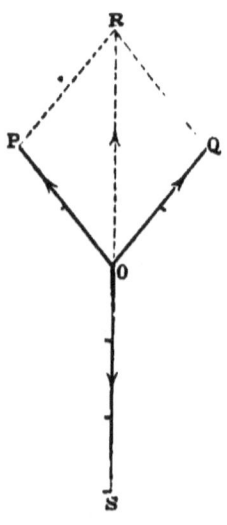

Fig. 5.

Let O P, O Q (Fig. 5) be the directions of the cords; O P and O Q being each of the length which corresponds to 1 lb., while O S corresponds to 1·5 lb. Here, as before, O P and O Q together may be considered to counteract O S. But O S could have been counteracted by an equal and opposite force O R. Hence O R may be regarded as the single force equivalent to O P and O Q, that is, as their resultant; and thus it is proved experimentally that these forces have a resultant. We can further verify that the resultant is the diagonal of the

I.]   EQUILIBRIUM OF THREE FORCES.   11

parallelogram of which the equal forces are the sides. Construct a parallelogram on a piece of cardboard having its four sides equal, and one of the diagonals half as long again as one of the sides. This may be done very easily by first drawing one of the two triangles into which the diagonal divides the parallelogram. The diagonal is to be produced beyond the parallelogram in the direction o s. When the cardboard is placed close against the cords, the two cords will lie in the directions o P, o Q, while the produced diagonal will be in the vertical o s. Thus the application of the parallelogram of force is verified.

16. The same experiment shows that two unequal forces may be compounded into one resultant. For in Fig. 5 the two forces o P and o s may be considered to be counterbalanced by the force o Q; in other words, o Q must be equal and opposite to a force which is the resultant of o P and o s.

17. Let us place on the central hook G a weight of 5 lbs., and weights of 3 lbs. on the hook E and 4 lbs. on F. This is actually the case shown in Fig. 3. The weights being unequal, we cannot immediately infer anything with reference to the position of the cords, but still we find, as before, that the cords assume a definite position, to which they return when temporarily displaced. Let Fig. 6 represent the positions of the cords. No two of the angles are in this case equal. Still each of the forces is counterbalanced by the other two. Each is therefore equal and opposite to the resultant of the other two. Construct

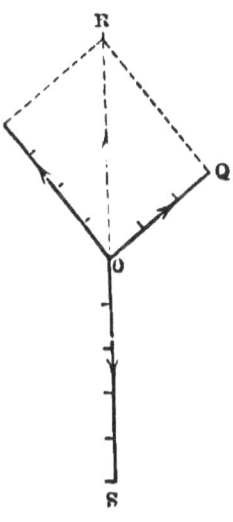

FIG. 6.

the parallelogram on cardboard, as can be easily done by forming the triangle O P R, whose sides are 3, 4, and 5, and then drawing O Q and R Q parallel to R P and O P. Produce the diagonal O R to S. This parallelogram being placed behind the cords, you see that the directions of the cords coincide with its sides and diagonal, thus verifying the parallelogram of forces in a case where all the forces are of different magnitudes.

18. It is easy, by the application of a set square, to prove that in this case the cords attached to the 3 lb. and 4 lb. weights are at right angles to each other. We could have inferred, from the parallelogram of force, that this must be the case, for the sides of the triangle O P R are 3, 4, and 5 respectively, and since the square of 5 is 25, and the squares of 3 and of 4 are 9 and 16 respectively, it follows that the square of one side of this triangle is equal to the sum of the squares of the two opposite sides, and therefore this is a right-angled triangle (Euclid, i. 48). Hence, since P R is parallel to O Q, the angle P O Q must also be a right angle.

A SMALL FORCE SOMETIMES BALANCES TWO LARGER FORCES.

19. Cases might be multiplied indefinitely by placing various amounts of weight on the hooks, constructing the parallelogram on cardboard, and comparing it with the cords as before. We shall, however, confine ourselves to one more illustration, which is capable of very remarkable applications. Attach 1 lb. to each of the hooks E and F; the cord joining them remains straight until drawn down by placing a weight on the centre hook. A very small weight will suffice to do this. Let us put on half-a-pound; the position the cords

I.] A SMALL FORCE BALANCING TWO LARGER 13

then assume is indicated in Fig. 7. As before, each force is equal and opposite to the resultant of the other two. Hence a force of half-a-pound is the resultant of two forces each of 1 lb. The apparent paradox is explained by noticing that the forces of 1 lb. are very nearly opposite, and therefore to a large extent counteract each other.

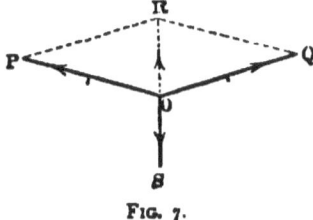

FIG. 7.

Constructing the cardboard parallelogram we may easily verify that the principle of the parallelogram of forces holds in this case also.

20. No matter how small be the weight we suspend from the middle of a horizontal cord, you see that the cord is deflected: and no matter how great a tension were applied, it would be impossible to straighten the cord. The cord could break, but it could not again become horizontal. Look at a telegraph wire; it is never in a straight line between two consecutive poles, and its curved form is more evident the greater be the distance between the poles. But in putting up a telegraph wire great straining force is used, by means of special machines for the purpose; yet the wires cannot be straightened: because the weight of the heavy wire itself acts as a force pulling it downwards. Just as the cord in our experiments cannot be straight when any force, however small, is pulling it downwards at the centre, so it is impossible by any exertion of force to straighten the long wire. Some further illustrations of this principle will be given in our next lecture, and with one application of it the present will be concluded.

21. One of the most important practical problems in mechanics is to make a small force overcome a greater. There are a number of ways in which this may be

accomplished for different purposes, and to the consideration of them several lectures of this course will be devoted. Perhaps, however, there is no arrangement more simple than that which is furnished by the principles we have been considering. We shall employ it to raise a 28 lb. weight by means of a 2 lb. weight. I do not say that this particular application is of much practical use. I show it to you rather as a remarkable deduction from the parallelogram of forces than as a useful machine.

A rope is attached at one end of an upright, A (Fig. 8),

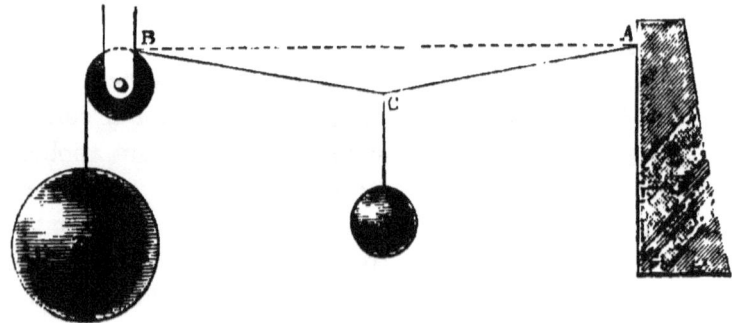

FIG. 8.

and passes over a pulley B at the same vertical height about 16' distant. A weight of 28 lbs. is fastened to the free end of the rope, and the supports must be heavily weighted, or otherwise secured from moving. The rope AB is apparently straight and horizontal, in consequence of its weight being inappreciable in comparison with the strain (28 lbs.) to which it is subjected; this position is indicated in the figure by the dotted line AB. We now suspend from C at the middle of the rope a weight of 2 lbs. Instantly the rope moves to the position represented in the figure. But this it cannot do without at the same moment raising slightly the 28 lbs., for, since two sides of a triangle, CB,

CA, are greater than the third side, AB, more of the rope must lie between the supports when it is bent down by the 2 lb. weight than when it was straight. But this can only have taken place by shortening the rope between the pulley B and the 28 lb. weight, for the rope is firmly secured at the other end. The effect on the heavy weight is so small that it is hardly visible to you from a distance. We can, however, easily show by an electrical arrangement that the big weight has been raised by the little one.

22. When an electric current passes through this alarum you hear the bell ring, and the moment I stop the current the bell stops. I have fastened one piece of brass to the 28 lb. weight, and another to the support close above it, but unless the weight be raised a little the two will not be in contact; the electricity is intended to pass from one of these pieces of brass to the other, but it cannot pass unless they are touching. When the rope is straight the two pieces of brass are separated, the current does not pass, and our alarum is dumb; but the moment I hang on the 2 lb. weight to the middle of the rope it raises the weight a little, brings the pieces of brass in contact, and now you all hear the alarum. On removing the 2 lbs. the current is interrupted and the noise ceases.

23. I am sure you must all have noticed that the 2 lb. weight descended through a distance of many inches, easily visible to all the room; that is to say, the small weight moved through a very considerable distance, while in so doing it only raised the larger one a very small distance. This is a point of the very greatest importance; I therefore take the first opportunity of calling your attention to it.

LECTURE II.

THE RESOLUTION OF FORCES.

Introduction.—One Force resolved into Two Forces. — Experimental Illustrations.— Sailing.—One Force resolved into Three Forces not in the same Plane.—The Jib and Tie-rod.

INTRODUCTION.

24. As the last lecture was principally concerned with discussing how one force could replace two forces, so in the present we shall examine the converse question, How may two forces replace one force? Since the diagonal of a parallelogram represents a single force equivalent to those represented by the sides, it is obvious that one force may be resolved into two others, provided it be the diagonal of the parallelogram formed by them.

25. We shall frequently employ in the present lecture, and in some of those that follow, the spring balance, which is represented in Fig. 9: the weight is attached to the hook, and when the balance is suspended by the

FIG. 9.

L. II.] ONE FORCE RESOLVED INTO TWO FORCES. 17

ring, a pointer indicates the number of pounds on a scale. This balance is very convenient for showing the strain along a cord; for this purpose the balance is held by the ring while the cord is attached to the hook. It will be noticed that the balance has two rings and two corresponding hooks. The hook and ring at the top and bottom will weigh up to 300 lbs., corresponding to the scale which is seen. The hook and ring at the side correspond to another scale on the other face of the plate: this second scale weighs up to about 50 lbs., consequently for a weight under 50 lbs. the side hook and ring are employed, as they give a more accurate result than would be obtained by the top and bottom hook and ring, which are intended for larger weights. These ingenious and useful balances are sufficiently accurate, and can easily be tested by raising known weights. Besides the instrument thus described, we shall sometimes use one of a smaller size, and we shall be able with this aid to trace the existence and magnitude of forces in a most convenient manner.

ONE FORCE RESOLVED INTO TWO FORCES.

26. We shall first illustrate how a single force may be resolved into a pair of forces; for this purpose we shall use the arrangement shown in Fig. 10 (see next page).

The ends of a cord are fastened to two small spring balances; to the centre E of this cord a weight of 4 lbs. is attached. At A and B are pegs from which the balances can be suspended. Let the distances AE, BE be each 12", and the distance AB 16". When the cord is thus placed, and the weight allowed to hang freely, each of the cords EA, EB is strained by an amount of force that is shown to be very nearly 3 lbs. by the balances. But the weight of 4 lbs. is the

C

only weight acting; hence it must be equivalent to two forces of very nearly 3 lbs. each along the directions AE and BE. Here the two forces to which 4 lbs. is equivalent are each of them less than 4 lbs., though taken together they exceed it.

27. But remove the cords from AB and hang them on CD, the length CD being 1' 10", then the forces shown along FC and FD are each 5 lbs.; here, therefore, one force of

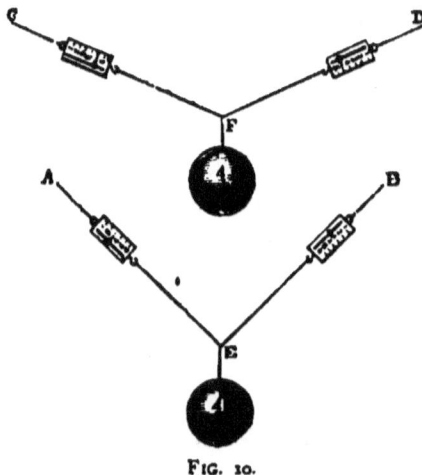

FIG. 10.

4 lbs. is equivalent to two forces each of 5 lbs. In the last lecture (Art. 19) we saw that one force could balance two greater forces; here we see the analogous case of one force being changed into two greater forces. Further, we learn that the number of pairs of forces into which one force may be decomposed is unlimited, for with every different distance between the pegs different forces will be indicated by the balances.

Whenever the weight is suspended from a point half-way between the balances, the forces along the cords are equal; but by placing the weight nearer one balance than the other, a greater force will be indicated on that balance to which the weight is nearest.

EXPERIMENTAL ILLUSTRATIONS.

28. The resolution or decomposition of one force into two forces each greater than itself is capable of being

FIG. 11.

illustrated in a variety of ways, two of which will be here explained. In Fig. 11 an arrangement for this purpose is shown. A piece of stout twine AB, able to support from 20 lbs. to 30 lbs., is fastened at one end A to a fixed support, and at the other end B to the eye of a wire-strainer. A wire-strainer consists of an iron rod, with an eye at one end and a screw and a nut at the other; it is used for tightening wires in wire fencing, and is employed in this case for the purpose of stretching the cord. This being done, I take a piece of ordinary sewing-thread, which is of course weaker than the stout twine. I tie the thread to the middle of the

FIG. 17.

cord at C, catch the other end in my fingers, and pull; something must break—something has broken: but what has broken? Not the slight thread, it is still whole; it is the cord which has snapped. Now this illustrates the point on which we have been dwelling. The force which I transmitted along the thread was insufficient to break it;

the thread transferred the force to the cord, but under such circumstances that the force was greatly magnified, and the consequence was that this magnified force was able to break the cord before the original force could break the thread. We can also see why it was necessary to stretch the cord. In Fig. 10 the strains along the cords are greater when the cords are attached at C and D than when they are attached at A and B; that is to say, the more the cord is stretched towards a straight line, the greater are the forces into which the applied force is resolved.

29. We give a second example, in illustration of the same principle.

In Fig. 12 is shown a chain 8' long, one end of which B is attached to a wire-strainer, while the other end is fastened to a small piece of pine A, which is $0''\cdot 5$ square in section, and 5" long between the two upright irons by which it is supported. By means of the nut of the wire-strainer I straighten the chain as I did the string of Fig. 11, and for the same reason. I then put a piece of twine round the chain and pull it gently. The strain brought to bear on the wood is so great that it breaks across. Here, the small force of a few pounds, transmitted to the chain by pulling the string, is magnified to upwards of a hundred-weight, for less than this would not break the wood. The explanation is precisely the same as when the string was broken by the thread.

SAILING.

30. The action of the wind upon the sails of a vessel affords a very instructive and useful example of the decomposition of forces. By the parallelogram of forces we are able to explain how it is that a vessel is able even to sail

against the wind. A force is that which tends to produce motion, and motion generally takes place in the line of the force. In the case of the action of wind on a vessel through the medium of the sails, we have motion produced which is not necessarily in the direction of the wind, and which may be to a certain extent opposed to it. This apparent paradox requires some elucidation.

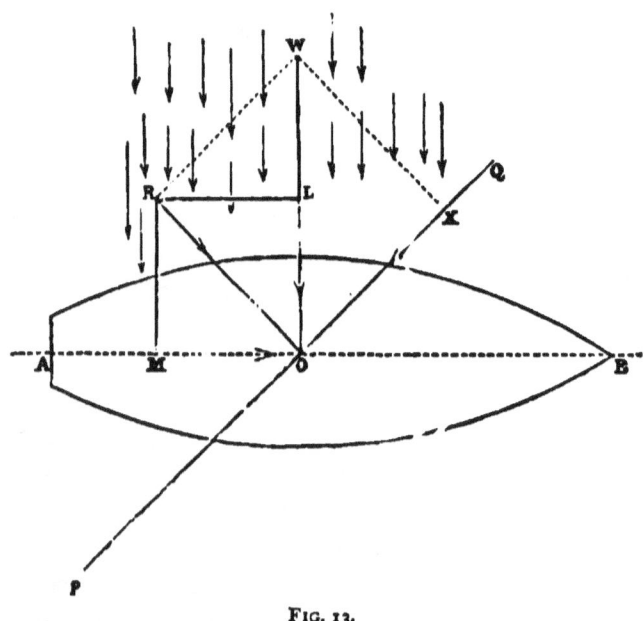

FIG. 13.

31. Let us first suppose the wind to be blowing in a direction shown by the arrows of Fig. 13, perpendicular to the line AB in which the ship's course lies.

In what direction must the sail be set ? It is clear that the sail must not be placed along the line AD, for then the only effect of the wind would be to blow the vessel sideways ; nor could the sail be placed with its edge to the wind, that is, along the line o w, for then the wind would merely glide along the sail without producing a propelling force. Let, then, the sail be placed between the two positions, as in the direction P Q. The line o w represents the magnitude of the force of the wind pressing on the sail.

We shall suppose for simplicity that the sail extends on both sides of o. Through o draw O R perpendicular to P Q, and from w let fall the perpendicular W X on P Q, and W R on O R. By the principle of the parallelogram of forces, the force o w may be decomposed into the two forces o X and O R, since these are the sides of the parallelogram of which o w, the force of the wind, is the diagonal. We may then leave o w out of consideration, and imagine the force of the wind to be replaced by the pair of forces o X and O R; but the force o X cannot produce an effect, it merely represents a force which glides along the surface of the sail, not one which pushes against it; so far as this component goes, the sail has its edge towards it, and therefore the force produces no effect. On the other hand, the sail is perpendicular to the force O R, and this is therefore the efficient component.

The force of the wind is thus measured by O R, both in magnitude and direction: this force represents the actual pressure on the mast produced by the sail, and from the mast communicated to the ship. Still O R is not in the direction in which the ship is sailing: we must again decompose the force in order to find its useful effect. This is done by drawing through R the lines R L and R M parallel to O A and O W, thus forming the parallelogram O M R L. Hence, by the parallelogram of forces, the force O R is equivalent to the two forces O L and O M.

The effect of O L upon the vessel is to propel it in a direction perpendicular to that in which it is sailing. We must, therefore, endeavour to counteract this force as far as possible. This is accomplished by the keel, and the form of the ship is so designed as to present the greatest possible resistance to being pushed sideways through the water: the deeper the keel the more completely is the effect of O L annulled. Still O L would in all cases produce some leeway were it not for the rudder, which, by turning the head of the vessel a little towards the wind, makes her sail in a direction sufficiently to windward to counteract the small effect of O L in driving her to leeward.

Thus O L is disposed of, and the only force remaining is O M, which acts directly to push the vessel in the required direction. Here, then, we see how the wind, aided by the resistance of the water, is able to make the vessel move in a direction perpendicular to that in which the wind blows. We have seen that the sail must be set somewhere between the direction of the wind and that of the ship's motion. It can be proved that when the direction of the sail supposed to be flat and vertical, is such as to bisect the angle W O B, the magnitude of the force O M is greater than when the sail has any other position.

32. The same principles show how a vessel is able to sail against the wind: she cannot, of course, sail straight against it, but she can sail within half a right angle of it, or perhaps even less. This can be seen from Fig. 14.

The small arrows represent the wind, as before. Let O W be the line parallel to them, which measures the force of the wind, and let the sail be placed along the line P Q; O W is decomposed into O X and O Y, O X merely glides along the sail, and O Y is the effective force. This is decomposed into O L and O M; O L is counteracted, as already explained, and O M is the force that propels the vessel onwards. Hence we see that there is a force acting to push the vessel onwards, even though the movement be partly against the wind.

It will be noticed in this case that the force O L acting to leewards exceeds O M pushing onwards. Hence it is that vessels with a very deep keel, and therefore opposing very great resistance to moving leewards, can sail more closely to the wind than others not so constructed; a vessel should be formed so that she shall move as freely as possible in the direction of her length, for which reason she is sharpened at the bow, and otherwise shaped for gliding through the water easily; this is in order that O M may have to overcome as

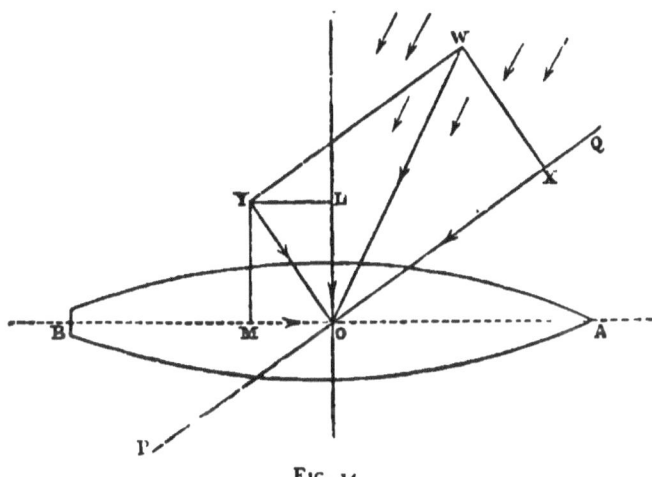

Fig. 14.

little resistance as possible. If the sail were flat and vertical it should bisect the angle A O W for the wind to act in the most efficient manner. Since, then, a vessel can sail towards the wind, it follows that, by taking a zigzag course, she can procced from one port to another, even though the wind be blowing from the place to which she would go towards the place from which she comes. This well-known manœuvre is called "tacking." You will understand that in a sailing-vessel the rudder has a more important part

26 EXPERIMENTAL MECHANICS. [LECT.

to play than in a steamer: in the latter it is only useful for changing the direction of the vessel's motion, while in the former it is not only necessary for changing the direction, but must also be used to keep the vessel to her course by counteracting the effect of leeway.

ONE FORCE RESOLVED INTO THREE FORCES NOT IN THE SAME PLANE.

33. Up to the present we have only been considering forces which lie in the same plane, but in nature we meet

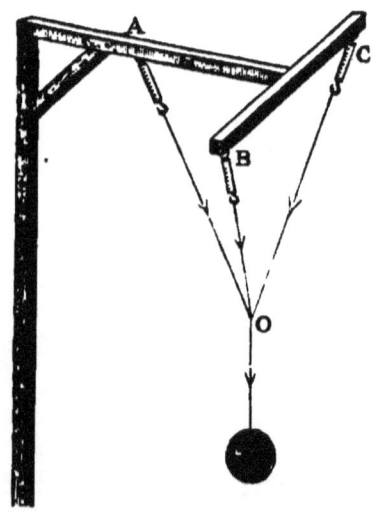

FIG. 15.

with forces acting in all directions, and therefore we must not be satisfied with confining our inquiries to the simpler case. We proceed to show, in two different ways, how a force can be decomposed into three forces not in the same plane, though passing through the same point. The first mode of doing so is as follows. To three points A, B, C

II.] ONE FORCE RESOLVED INTO THREE. 27

(Fig. 15) three spring balances are attached; A, B, C are not in the same straight line, though they are at the same vertical height: to the spring balances cords are attached, which unite in a point O, from which a weight W is suspended. This weight is supported by the three cords, and the strains along these cords are indicated by the spring balances. The greatest strain is on the shortest cord and the least strain on the longest. Here the force W lbs. produces three forces which, taken together, exceed its own amount. If I add an equal weight W, I find, as we might have anticipated, that the strains indicated by the scales are precisely double what they were before. Thus we see that the proportion of the force to each of the components into which it is decomposed does not depend on the actual magnitude of the force, but on the relative direction of the force and its components.

34. Another mode of showing the decomposition of one force into three forces not in the same plane is represented in Fig. 16. The tripod is formed of three strips of pine, 4' × 0"·5 × 0"·5, secured by a piece of wire running through each at the top; one end of this wire hangs down, and carries a hook to which is attached a weight of 28 lbs. This weight is supported by the wire, but the strain on the wire must be borne by the three wooden rods: hence there is a force acting downwards through the wooden rods. We cannot render this manifest by a contrivance like

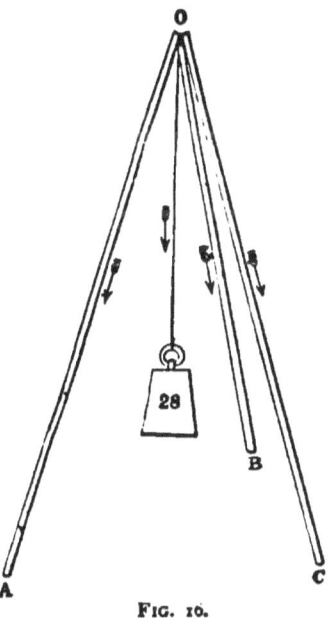

FIG. 16.

the spring scales, because it is a push instead of a pull. However, by raising one of the legs I at once become aware that there is a force acting downwards through it. The weight is, then, decomposed into three forces, which act downwards through the legs; these three forces are not in a plane, and the three forces taken together are larger than the weight.

35. The tripod is often used for supporting weights; it is convenient on account of its portability, and it is very steady. You may judge of its strength by the model represented in the figure, for though the legs are very slight, yet they support very securely a considerable weight. The pulleys by means of which gigantic weights are raised are often supported by colossal tripods. They possess stability and steadiness in addition to great strength.

36. An important point may be brought out by contrasting the arrangements of Figs. 15 and 16. In the one case three cords are used, and in the other three rods. Three rods would have answered for both, but three cords would not have done for the tripod. In one the cords are strained, and the tendency of the strain is to break the cords, but in the other the nature of the force down the rods is entirely different; it does not tend to pull the rod asunder, it is trying to crush the rod, and had the weight been large enough the rods would bend and break. I hold one end of a pencil in each hand and then try to pull the pencil asunder; the pencil is in the condition of the cords of Fig. 15; but if instead of pulling I push my hands together, the pencil is like the rods in Fig. 16.

37. This distinction is of great importance in mechanics. A rod or cord is in a state of tension is called a "tie"; while a rod in a state of compression is called a "strut." Since a rod can resist both tension and compression it can serve

THE JIB AND TIE ROD.

either as a tie or as a strut, but a cord or chain can only act as a tie. A pillar is always a strut, as the superincumbent load makes it to be in a state of compression. These distinctions will be very frequently used during this course of lectures, and it is necessary that they be thoroughly understood.

THE JIB AND TIE ROD.

38. As an illustration of the nature of the "tie" and "strut," and also for the purpose of giving a useful example of the decomposition of forces, I use the apparatus of Fig. 17 (see next page).

It represents the principle of the framework in the common lifting crane, and has numerous applications in practical mechanics. A rod of wood BC 3' 6" long and 1" × 1" section is capable of turning round its support at the bottom B by means of a joint or hinge: this rod is called the "jib"; it is held at its upper end by a tie AC 3' long, which is attached to the support above the joint. AB is one foot long. From the point C a wire descends, having a hook at the end on which a weight can be hung. The tie is attached to the spring balance, the index of which shows the strain. The spring balance is secured by a wire-strainer, by turning the nut of which the length of the wire can be shortened or lengthened as occasion requires. This is necessary, because when different weights are suspended from the hook the spring is stretched more or less, and the screw is then employed to keep the entire length of the tie at 3'. The remainder of the tie consists of copper wire.

39. Suppose a weight of 20 lbs. be suspended from the hook W, it endeavours to pull the top of the jib downwards; but the tie holds it back, consequently the tie is put into a

state of tension, as indeed its name signifies, and the magnitude of that tension is shown to be 60 lbs. by the spring-balance. Here we find again what we have already so often referred to; namely, one force developing another force that is greater than itself, for the strain along the tie is three times

Fig. 17.

as great as the strain in the vertical wire by which it was produced.

40. What is the condition of the jib? It is evidently being pushed downwards on its joint at B; it is therefore in a state of compression; it is a strut. This will be evident if we think for a moment how absurd it would be to

THE JIB AND TIE ROD.

deavour to replace the jib by a string or chain: the whole arrangement would collapse. The weight of 20 lbs. is therefore decomposed by this contrivance into two other forces, one of which is resisted by a tie and the other by a strut.

41. We have no means of showing the magnitude of the strain along the strut, but we shall prove that it can be computed by means of the parallelogram of force; this will also explain how it is that the tie is strained by a force three times that of the weight which is used. Through C (Fig. 18) draw C P parallel to the tie A B, and P Q parallel to the strut

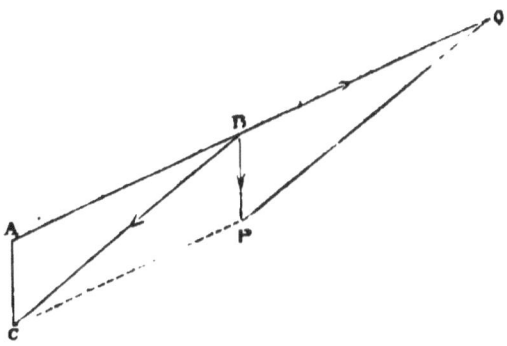

FIG. 18.

C B then B P is the diagonal of the parallelogram whose sides are each equal to B C and B Q. If therefore we consider the force of 20 lbs. to be represented by B P, the two forces into which it is decomposed will be shown by B Q and B C; but A B is equal to B Q, since each of them is equal to C P; also B P is equal to A C. Hence the weight of 20 lbs. being represented by A C, the strain along the tie will be represented by the length A B, and that along the strut by the length B C. Remembering that A B is 3' long, C B 3' 6", and A C 1', it follows that the strain along the tie is 60 lbs., and along the strut 70 lbs., when the weight of 20 lbs. is suspended from the hook.

42. In every other case the strains along the tie and strut can be determined, when the suspended weight is known, by their proportionality to the sides of the triangle formed by the tie, the jib, and the upright post, respectively.

43. In this contrivance you will recognize, no doubt, the framework of the common lifting crane, but that very essential portion of the crane which provides for the raising and lowering is not shown here. To this we shall return again in a subsequent lecture (Art. 332). You will of course understand that the tie rod we have been considering is entirely different from the chain for raising the load.

44. It is easy to see of what importance to the engineer the information acquired by means of the decomposition of forces may become. Thus in the simple case with which we are at present engaged, suppose an engineer were required to erect a frame which was to sustain a weight of 10 tons, let us see how he would be enabled to determine the strength of the tie and jib. It is of importance in designing any structure not to make any part unnecessarily strong, as doing so involves a waste of valuable material, but it is of still more vital importance to make every part strong enough to avoid the risk of accident, not only under ordinary circumstances, but also under the exceptionally great shocks and strains to which every machine is liable.

45. According to the numerical proportions we have employed for illustration, the strain along the tie rod would be 30 tons when the load was 10 tons, and therefore the tie must at least be strong enough to bear a pull of 30 tons; but it is customary, in good engineering practice, to make the machine of about ten times the strength that would just be sufficient to sustain the ordinary load. Hence the crank

must be so strong that the tie would not break with a tension less than 300 tons, which would be produced when the crane was lifting 100 tons. So great a margin of safety is necessary on account of the jerks and other occasional great strains that arise in the raising and the lowering of heavy weights. For a crane intended to raise 10 tons, the engineer must therefore design a tie rod which not less than 300 tons would tear asunder. It has been proved by actual trial that a rod of wrought iron of average quality, one square inch in section, can just withstand a pull of twenty tons. Hence fifteen such rods, or one rod the section of which was equal to fifteen square inches, would be just able to resist 300 tons; and this is therefore the proper area of section for the tie rod of the crane we have been considering.

46. In the same way we ascertain the actual thrust down the jib; it amounts to 35 tons, and the jib should be ten times as strong as a strut which would collapse under a strain of 35 tons.

47. It is easy to see from the figure that the tie rod is pulling the upright, and tending, in fact, to make it snap off near B. It is therefore necessary that the upright support A B (Fig. 17) be secured very firmly.

LECTURE III.

PARALLEL FORCES.

Introduction.—Pressure of a Loaded Beam on its Supports.—Equilibrium of a Bar supported on a Knife-edge.—The Composition of Parallel Forces.—Parallel Forces acting in opposite directions.—The Couple.—The Weighing Scales.

INTRODUCTION.

48. THE parallelogram of forces enables us to find the resultant of two forces which intersect: but since parallel forces do not intersect, the construction does not avail to determine the resultant of two parallel forces. We can, however, find this resultant very simply by other means.

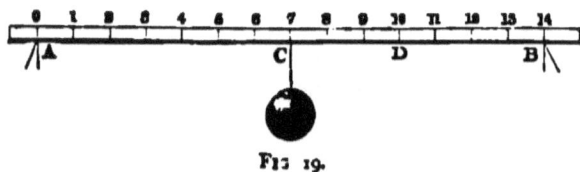

Fig 19.

49. Fig. 19 represents a wooden rod 4' long, sustained by resting on two supports A and B, and having the length A B divided into 14 equal parts. Let a weight of 14 lbs. be hung on the rod at its middle point C; this weight must be borne by the supports, and it is evident that they will bear

LECT. III.] PRESSURE OF A LOADED BEAM. 35

the load in equal shares, for since the weight is at the middle of the rod there is no reason why one end should be differently circumstanced from the other. Hence the total pressure on each of the supports will be 7 lbs., together with half the weight of the wooden bar.

50. If the weight of 14 lbs. be placed, not at the centre of the bar, but at some other point such as D, it is not then so easy to see in what proportion the weight is distributed between the supports. We can easily understand that the support near the weight must bear more than the remote one, but how much more? When we are able to answer this question, we shall see that it will lead us to a knowledge of the composition of parallel forces.

PRESSURE OF A LOADED BEAM ON ITS SUPPORTS.

51. To study this question we shall employ the apparatus shown in Fig. 20. An iron bar 5' 6" long, weighing 10 lbs., rests in the hooks of the spring balances A,C, in the manner shown in the figure. These hooks are exactly five feet apart, so that the bar projects 3" beyond each end. The space between the hooks is divided into twenty equal portions, each of course 3" long. The bar is sufficiently strong to bear the weight B of 20 lbs. suspended from it by an S hook, without appreciable deflection. Before the weight of 20 lbs. is suspended, the spring balances each show a strain of 5 lbs. We would expect this, for it is evident that the whole weight of the bar amounting to 10 lbs. should be borne equally by the two supports.

52. When I place the weight in the middle, 10 divisions from each end, I find the balances each indicate 15 lbs. But 5 lbs. is due to the weight of the bar. Hence the 20 lbs. is divided equally, as we have already stated that it

D 2

should be. But let the 20 lbs. be moved to any other position, suppose 4 divisions from the right, and 16 from the left; then the right-hand scale reads 21 lbs., and the left-hand reads 9 lbs. To get rid of the weight of the bar itself, we must subtract 5 lbs. from each. We learn therefore that the 20 lb. weight pulls the right-hand spring balance with a strain of 16 lbs., and the left with a strain of

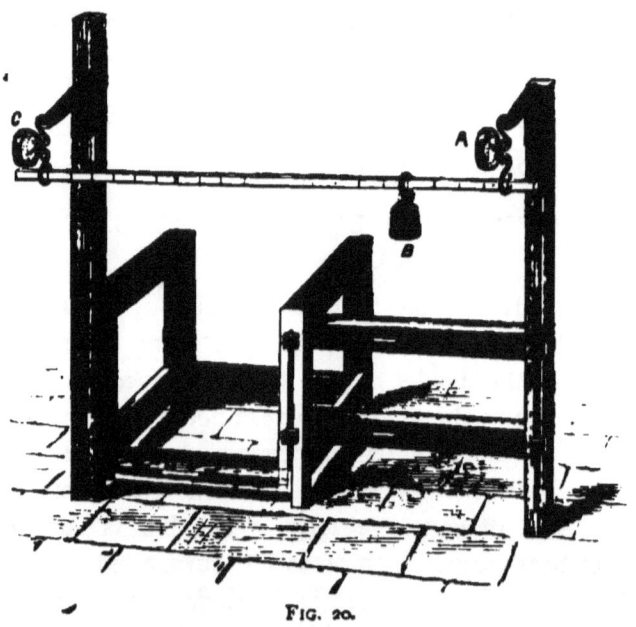

FIG. 20.

4 lbs. Observe this closely; you see I have made the number of divisions in the bar equal to the number of pounds weight suspended from it, and here we find that when the weight is 16 divisions from the left, the strain of 16 lbs. is shown on the right. At the same time the weight is 4 divisions from the right, and 4 lbs. is the strain shown to the left.

PRESSURE OF A LOADED BEAM.

53. I will state the law of the distribution of the load a little more generally, and we shall find that the bar will prove the law to be true in all cases. *Divide the bar into as many equal parts as there are pounds in the load, then the pressure in pounds on one end is the number of divisions that the load is distant from the other.*

54. For example, suppose I place the load 2 divisions from one end: I read by the scale at that end 23 lbs.; subtracting 5 lbs. for the weight of the bar, the pressure due to the load is shown to be 18 lbs., but the weight is then exactly 18 divisions distant from the other end. We can easily verify this rule whatever be the position which the load occupies.

55. If the load be placed between two marks, instead of being, as we have hitherto supposed, exactly at one, the partition of the load is also determined by the law. Were it, for example, 3·5 divisions from one end, the strain on the other would be 3·5 lbs.; and in like manner for other cases.

56. We have thus proved by actual experiment this useful and instructive law of nature; the same result could have been inferred by reasoning from the parallelogram of force, but the purely experimental proof is more in accordance with our scheme. The doctrine of the composition of parallel forces is one of the most fundamental parts of mechanics, and we shall have many occasions to employ it in this as well as in subsequent lectures.

57. Returning now to Fig. 19, with which we commenced, the law we have discovered will enable us to find how the weight is distributed. We divide the length of the bar between the supports into 14 equal parts because the weight is 14 lbs.; if, then, the weight be at D, 10 divisions from one end A, and 4 from the other B, the

pressure at the corresponding ends will be 4 and 10. If the weight were 2·5 divisions from one end, and therefore 11·5 from the other, the shares in which this load would be supported at the ends are 11·5 lbs. and 2·5 lbs. The actual pressure sustained by each end is, however, about 6 ounces greater if the weight of the wooden bar itself be taken into account.

58. Let us suspend a second weight from another point of the bar. We must then calculate the pressures at the ends which each weight separately would produce, and those at the same end are to be added together, and to half the weight of the bar, to find the total pressure. Thus, if one weight of 20 lbs. were in the middle, and another of 14 lbs. at a distance of 11 divisions from one end, the middle weight would produce 10 lbs. at each end and the 14 lbs. would produce 3 lbs. and 11 lbs., and remembering the weight of the bar, the total pressures produced would be 13 lbs. 6 oz. and 21 lbs. 6 oz. The same principles will evidently apply to the case of several weights: and the application of the rule becomes especially easy when all the weights are equal, for then the same divisions will serve for calculating the effect of each weight.

59. The principles involved in these calculations are of so much importance that we shall further examine them by a different method, which has many useful applications.

EQUILIBRIUM OF A BAR SUPPORTED ON A KNIFE-EDGE.

60. The weight of the bar has hitherto somewhat complicated our calculations; the results would appear more simply if we could avoid this weight; but since we want a strong bar, its weight is not so small that we could afford to

III.] EQUILIBRIUM OF A BAR. 39

overlook it altogether. By means of the arrangement of

Fig. 21.

Fig. 21, we can *counterpoise* the weight of the bar. To

the centre of A B a cord is attached, which, passing over a fixed pulley D, carries a hook at the other end. The bar, being a pine rod, 4 feet long and 1 inch square, weighs about 12 ounces; consequently, if a weight of twelve ounces be suspended from the hook, the bar will be counterpoised, and will remain at whatever height it is placed.

61. A B is divided by lines drawn along it at distances of 1″ apart; there are thus 48 of these divisions. The weights employed are furnished with rings large enough to enable them to be slipped on the bar and thus placed in any desired position.

62. Underneath the bar lies an important portion of the arrangement; namely, the knife-edge C. This is a blunt edge of steel firmly fastened to the support which carries it. This support can be moved along underneath the bar so that the knife-edge can be placed under any of the divisions required. The bar being counterpoised, though still unloaded with weights, may be brought down till it just touches the knife-edge; it will then remain horizontal, and will retain this position whether the knife-edge be at either end of the bar or in any intermediate position. I shall hang weights at the extremities of the rod, and we shall find that there is for each pair of weights just one position at which, if the knife-edge be placed, it will sustain the rod horizontally. We shall then examine the relations between these distances and the weights that have been attached, and we shall trace the connection between the results of this method and those of the arrangement that we last used.

63. Supposing that 6 lbs. be hung at each end of the rod, we might easily foresee that the knife-edge should be placed in the middle, and we find our anticipations verified. When the edge is exactly at the middle, the rod remains horizontal; but if it be moved, even through a very small

EQUILIBRIUM OF A BAR.

distance, to either side, the rod instantly descends on the other. The knife-edge is 24 inches distant from each end; and if I multiply this number by the number of pounds in the weight, in this case 6, I find 144 for the product, and this product is the same for both ends of the bar. The importance of this remark will be seen directly.

64. If I remove one of the 6 lb. weights and replace it by 2 lbs., leaving the other weight and the knife-edge unaltered, the bar instantly descends on he side of the heavy weight; but, by slipping the knife-edge along the bar, I find that when I have moved it to within a distance of 12 inches from the 6 lbs., and therefore 36 inches from the 2 lbs., the bar will remain horizontal. The edge must be put carefully at the right place; a quarter of an inch to one side or the other would upset the bar. The whole load borne by the knife-edge is of course 8 lbs., being the sum of the weights. If we multiply 2, the number of pounds at one end, by 36, the distance of that end from the knife-edge, we obtain the product 72; and we find precisely the same product by multiplying 6, the number of pounds in the other weight, by 12, its distance from the knife-edge. To express this result concisely we shall introduce the word *moment*, a term of frequent use in mechanics. The 2 lb. weight produces a force tending to pull its end of the bar downwards by making the bar turn round the knife-edge. *The magnitude of th's force, multiplied into its distance from the knife-edge, is called the moment of the force.* We can express the result at which we have arrived by saying that, when the knife-edge has been so placed that the bar remains horizontal, *the moments of the forces about the knife-edge are equal.*

65. We may further illustrate this law by suspending weights of 7 lbs. and 5 lbs. respectively from the ends of

the bar; it is found that the knife-edge must then be placed 20 inches from the larger weight, and, therefore, 28 inches from the smaller, but $5 \times 28 = 140$, and $7 \times 20 = 140$, thus again verifying the law of equality of the moments.

From the equality of the moments we can also deduce the law for the distribution of the load given in Art. 53. Thus, taking the figures in the last experiment, we have loads of 7 lbs. and 5 lbs. respectively. These produce a pressure of $7 + 5 = 12$ lbs. on the knife-edge. This edge presses on the bar with an equal and opposite reaction. To ascertain the distribution of this pressure on the ends of the beam, we divide the whole beam into 12 equal parts of 4 inches each, and the 7 lb. weight is 5 of these parts, *i.e.*, 20 inches distant from the support. Hence the edge should be 20 inches from the greater weight, which is the condition also implied by the equality of the moments.

THE COMPOSITION OF PARALLEL FORCES.

66. Having now examined the subject experimentally, we proceed to investigate what may be learned from the results we have proved.

The weight of the bar being allowed for in the way we have explained, by subtracting one half of it from each of the strains indicated by the spring balance (FIG. 20), we may omit it from consideration. As the balances are pulled downwards by the bar when it is loaded, so they will react to pull the bar upwards. This will be evident if we think of a weight—say 14 lbs.—suspended from one of these balances : it hangs at rest ; therefore its weight, which is constantly urging it downwards, must be counteracted by an equal force pulling it upwards. The balance of course shows 14 lbs. ; thus the spring exerts in an upward pull a force which is precisely equal to that by which it is itself pulled downwards.

67. Hence the springs are exerting forces at the ends of the bar in pulling them upwards, and the scales indicate their magnitudes. The bar is thus subject to three forces, viz.: the suspended weight of 20 lbs., which acts vertically downwards, and the two other forces which act vertically upwards, and the united action of the three make equilibrium.

68. Let lines be drawn, representing the forces in the manner already explained. We have then three parallel forces AP, BQ, CR acting on a rod in equilibrium (Fig. 22). The two forces AP and BQ may be considered as balanced by the force CR in the position shown in the figure, but the force CR would be balanced by the equal and opposite force CS, represented by the dotted line. Hence this last force is equivalent to AP and BQ. In other words, it must be their resultant. Here then we learn that a pair of parallel forces, acting in the same direction, can be compounded into a single resultant.

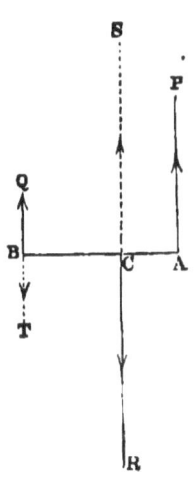

FIG. 22.

69. We also see that the magnitude of the resultant is equal to the sum of the magnitudes of the forces, and further we find the position of the resultant by the following rule. Add the two forces together; divide the distance between them into as many equal parts as are contained in the sum, measure off from the greater of these two forces as many parts as there are pounds in the smaller force, and that is the point required. This rule is very easily inferred from that which we were taught by the experiments in Art. 51.

PARALLEL FORCES ACTING IN OPPOSITE DIRECTIONS.

70. Since the forces AP, BQ, CR (Fig. 22) are in equilibrium, it follows that we may look on BQ as balancing in the position which it occupies the two forces of AP and CR in their positions. This may remind us of the numerous instances we have already met with, where one force balanced two greater forces: in the present case AP and CR are acting in opposite directions, and the force BQ which balances them is equal to their difference. A force BT equal and opposite to BQ must then be the resultant of CR and AP, since it is able to produce the same effect. Notice that in this case the resultant of the two forces is not between them, but that it lies on the side of the larger. When the forces act in the same direction, the resultant is always between them.

71. The actual position which the resultant of two opposite parallel forces occupies is to be found by the following rule. Divide the distance between the forces into as many equal parts as there are pounds in their difference, then measure outwards from the point of application of the larger force as many of these parts as there are pounds in the smaller force; the point thus found determines the position of the resultant. Thus, if the forces be 14 and 20, the difference between them is 6, and therefore the distance between their directions is divided into six parts; from the point of application of the force of 20, 14 parts are measured outwards, and thus the position of the resultant is determined. Hence we have the means of compounding two parallel forces in general

THE COUPLE.

72. In one case, however, two parallel forces have no resultant; this occurs when the two forces are equal, and in

THE COUPLE.

opposite directions. A pair of forces of this kind is called *a couple;* there is no single force which could balance a couple,—it can only be counterbalanced by another couple acting in an opposite manner. This remarkable case, may be studied by the arrangement of Fig. 23.

A wooden rod, A B 48″ × 0″·5 × 0″·5, has strings attached to it at points A and D, one foot distant. The string at D passes over a fixed pulley E, and at the end P a hook is attached for the purpose of receiving weights, while a similar hook depends from A; the weight of the rod

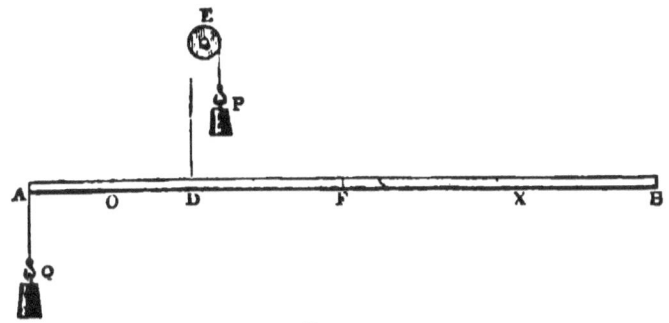

FIG. 23.

itself, which only amounts to three ounces, may be neglected, as it is very small compared with the weights which will be used.

73. Supposing 2 lbs. to be placed at P, and 1 lb. at Q, we have two parallel forces acting in opposite directions; and since their difference is 1 lb., it follows from our rule that the point F, where D F is equal to A D, is the point where the resultant is applied. You see this is easily verified, for by placing my finger over the rod at F it remains horizontal and in equilibrium; whereas, when I move my finger to one side or the other, equilibrium is impossible. If I move it nearer to B, the end A ascends. If I move it towards A, the end B ascends.

74. To study the case when the two forces are equal, a load of 2 lbs. may be placed on each of the hooks P and Q. It will then be found that the finger cannot be placed in any position along the rod so as to keep it in equilibrium; that is to say, no single force can counteract the two forces which form the couple. Let O be the point midway between A and D. The forces evidently tend to raise O B and turn the part O A downwards; but if I try to restrain O B by holding my finger above, as at the point X, instantly the rod begins to turn round X and the part from A to X descends. I find similarly that any attempt to prevent the motion by holding my finger underneath is equally unsuccessful. But if at the same time I press the rod downwards at one point, and upwards at another with suitable force, I can succeed in producing equilibrium; in this case the two pressures form a couple; and it is this couple which neutralizes the couple produced by the weights. We learn, then, the important result that *a couple can be balanced by a couple*, and by a couple only.

75. The moment of a couple is the product of one of the two equal forces into their perpendicular distance. Two couples tending to turn the body to which they are applied in the *same* direction will be equivalent if their moments are equal. Two couples which tend to turn the body in *opposite* directions will be in equilibrium if their moments are equal. We can also compound two couples in the same or in opposite directions into a single couple of which the moment is respectively either the sum or the difference of the original moments.

THE WEIGHING SCALES.

76. Another apparatus by which the nature of parallel forces may be investigated is shown in Fig. 24; it con-

sists of a slight frame of wood A B C, 4' long. At E, a pair of steel knife-edges is clamped to the frame. The knife-edges rest on two pieces of steel, one of which is shown at O F. When the knife-edges are suitably placed the frame is balanced, so that a small piece of paper laid at A will cause that side to descend.

77. We suspend two small hooks from the points A and B : these are made of fine wire, so that their weight may be

FIG. 24.

left out of consideration. With this apparatus we can in the first place verify the principle of equality of moments : for example, if I place the hook A at a distance of 9" from the centre O and load it with 1 lb., I find that when B is laden with 0·5 lb. it must be at a distance of 18" from O in order to counterbalance A ; the moment in the one case is 9 × 1, in the other 18 × 0·5, and these are obviously equal.

78. Let a weight of 1 lb. be placed on each of the hooks, the frame will only be in equilibrium when the

hooks are at precisely the same distance from the centre. A familiar application of this principle is found in the ordinary weighing scales; the frame, which in this case is called a *beam*, is sustained by two knife-edges, smaller, however, than those represented in the figure. The pans P, P are suspended from the extremities of the beam, and should be at equal distances from its centre. These scale-pans must be of equal weight, and then, when equal weights are placed in them, the beam will remain horizontal. If the weight in one slightly exceed that in the other, the pan containing the heavier weight will of course descend.

79. That a pair of scales should weigh accurately, it is necessary that the weights be correct; but even with correct weights, a balance of defective construction will give an inaccurate result. The error f equently arises from some inequality in the lengths of the arms of the beam. When this is the case, the two weights which really balance are not equal. Supposing, for instance, that with an imperfect balance I endeavour to weigh a pound of shot. If I put the weight on the short side, then the quantity of shot balanced is less than 1 lb.; while if the 1 lb. weight be placed at the long side, it will require more than 1 lb. of shot to produce equilibrium. The mode of testing a pair of scales is then evident. Let weights be placed in the pans which balance each other; if the weights be interchanged and the balance still remains horizontal, it is correct.

80. Suppose, for example, that the two arms be 10 inches and 11 inches long, then, if 1 lb. weight be placed in the pan of the 10-inch end, its moment is 10; and if $\frac{10}{11}$ of 1 lb. be placed in the pan belonging to the 11-inch end, its moment is also 10: hence 1 lb. at the short end balances $\frac{10}{11}$ of 1 lb. at the long end; and therefore, if the shopkeeper placed his weight in the short arm, his customers would lose $\frac{1}{11}$

part of each pound for which they paid; on the other hand, if the shopkeeper placed his 1 lb. weight on the long arm, then not less than $1\frac{1}{10}$ lb. would be required in the pan belonging to the short arm. Hence in this case the customer would get $\frac{1}{10}$ lb. too much. It follows, that if a shopman placed his weights and his goods alternately in the one scale and in the other he would be a loser on the whole; for, though every second customer gets $\frac{1}{11}$ lb. less than he ought, yet the others get $\frac{1}{10}$ lb. more than they have paid for.

LECTURE IV.

THE FORCE OF GRAVITY.

Introduction.—Specific Gravity.—The Plummet and Spirit Level.—The Centre of Gravity.—Stable and Unstable Equilibrium.—Property of the Centre of Gravity in a Revolving Wheel.

INTRODUCTION.

81. IN the last three lectures we considered forces in the abstract; we saw how they are to be represented by straight lines, how compounded together and how decomposed into others; we have explained what is meant by forces being in equilibrium, and we have shown instances where the forces lie in the same plane or in different planes, and where they intersect or are parallel to each other. These subjects are the elements of mechanics; they form the framework which in this and subsequent lectures we shall try to present in a more attractive garb. We shall commence by studying the most remarkable force in nature, a force constantly in action, and one to which all bodies are subject, a force which distance cannot annihilate, and one the properties of which have led to the most sublime discoveries of human intellect. This is the force of gravity.

THE FORCE OF GRAVITY.

82. If I drop a stone from my hand, it falls to the ground. That which produces motion is a force: hence the stone must have been acted upon by a force which drew it to the ground. On every part of the earth's surface experience shows that a body tends to fall. This fact proves that there is an attractive force in the earth tending to draw all bodies towards it.

83. Let A B C D (Fig. 25) be points from which stones are let fall, and let the circle represent the section of the earth;

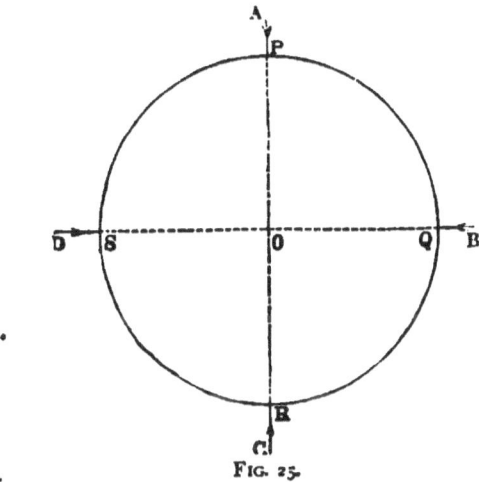

FIG. 25.

let P Q R S be the points at the surface of the earth upon which the stones will drop when allowed to do so. The four stones will move in the directions of the arrows: from A to P the stone moves in an opposite direction to the motion from C to R; from B to Q it moves from right to left, while from D to S it moves from left to right. The movements are in different directions; but if I produce these directions, as indicated by the dotted lines, they each pass through the centre O.

84. Hence each stone in falling moves towards the centre of the earth, and this is consequently the direction of the force. We therefore assert that the earth has an attraction for the stone, in consequence of which it tries to get as near the earth's centre as possible, and this attraction is called the force of gravitation.

85. We are so excessively familiar with the phenomenon of seeing bodies fall that it does not excite our astonishment or arouse our curiosity. A clap of thunder, which every one notices, because much less frequent, is not really more remarkable. We often look with attention at the attraction of a piece of iron by a magnet, and justly so, for the phenomenon is very interesting, and yet the falling of a stone is produced by a far grander and more important force than the force of magnetism.

86. It is gravity which causes the weight of bodies. I hold a piece of lead in my hand: gravity tends to pull it downwards, thus producing a pressure on my hand which I call *weight*. Gravity acts with slightly varying intensity at various parts of the earth's surface. This is due to two distinct causes, one of which may be mentioned here, while the other will be subsequently referred to. The earth is not perfectly spherical; it is flattened a little at the poles; consequently a body at the pole is nearer the general mass of the earth than a body at the equator; therefore the body at the pole is more attracted, and seems heavier. A mass which weighs 200 lbs. at the equator would weigh one pound more at the pole: about one-third of this increase is due to the cause here pointed out. (See Lecture XVII.)

87. Gravity is a force which attracts every particle of matter; it acts not merely on those parts of a body which lie on the surface, but it equally affects those in the interior. This is proved by observing that a body has the same

weight, however its shape be altered: for example, suppose I take a ball of putty which weighs 1 lb., I shall find that its weight remains unchanged when the ball is flattened into a thin plate, though in the latter case the surface, and therefore the number of superficial particles, is larger than it was in the former.

SPECIFIC GRAVITY.

88. Gravity produces different effects upon different substances. This is commonly expressed by saying that some substances are heavier than others; for example, I have here a piece of wood and a piece of lead of equal bulk. The lead is drawn to the earth with a greater force than the wood. Substances are usually termed heavy when they sink in water, and light when they float upon it. But a body sinks in water if it weigh more than an equal bulk of water, and floats if it weigh less. Hence it is natural to take water as a standard with which the weights of other substances may be compared.

89. I take a certain volume, say a cubic inch of cast iron such as this I hold in my hand, and which has been accurately shaped for the purpose. This cube is heavier than one cubic inch of water, but I shall find that a certain quantity of water is equal to it in weight; that is to say, a certain number of cubic inches of water, and it may be fractional parts of a cubic inch, are precisely of the same weight. This number is called the *specific gravity of cast iron*.

90. It would be impossible to counterpoise water with the iron without holding the water in a vessel, and the weight of the vessel must then be allowed for. I adopt the following plan. I have here a number of inch cubes of wood (Fig. 26), which would each be lighter than a cubic inch of water, but I have weighted the wooden cubes by placing

grains of shot into holes bored into the wood. The weight of each cube has thus been accurately adjusted to be equal to that of a cubic inch of water. This may be tested by actual weighing. I weigh one of the cubes and find it to be 252 grains, which is well known to be the weight of a cubic inch of water.

91. But the cubes may be shown to be identical in weight

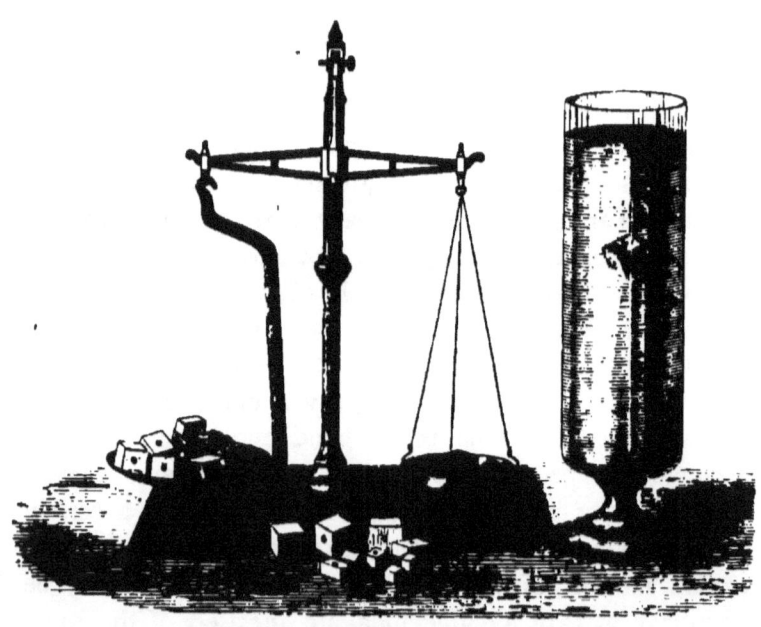

FIG. 26.

with the same bulk of water by a simpler method. One of them placed in water should have no tendency to sink, since it is not heavier than water, nor on the other hand, since it is not lighter, should it have any tendency to float. It should then remain in the water in whatever position it may be placed. It is difficult to prepare one of these cubes so accurately that this result should be attained, and it is impossible to ensure its continuance for any time owing to changes of temperature and the absorption of water by the wood. We can, however, by a slight modification, prove that one of these cubes is at all events nearly equal in weight to the same bulk of water. In Fig. 26 is shown a tall glass jar filled with a fluid in appearance like plain water, but it is really composed in the following manner. I first poured into the jar a very weak solution of salt and water, which partially filled it; I then poured gently upon this a little pure water, and finally filled up the jar with water containing a little spirits of wine : the salt and water is a little heavier than pure water, while the spirit and water is a little lighter. I take one of the cubes and drop it gently into the glass; it falls through the spirit and water, and after making a few oscillations settles itself at rest in the stratum shown in the figure. This shows that our prepared cube is a little heavier than spirit and water, and a little lighter than salt and water, and hence we infer that it must at all events be very near the weight of pure water which lies between the two. We have also a number of half cubes, quarter cubes, and half-quarter cubes, which have been similarly prepared to be of equal weight with an equal bulk of water.

92. We shall now be able to measure the specific gravity of a substance. In one pan of the scales I place the inch cube of cast iron, and I find that $7\frac{1}{4}$ of the wooden cubes, which we may call cubes of water, will balance it. We therefore say that the specific gravity of iron is $7\frac{1}{4}$. The exact number found by more accurate methods is 7·2. It is often convenient to remember that 23 cubic inches of cast iron weigh 6 lbs., and that therefore one cubic inch weighs very nearly $\frac{1}{4}$ lb.

93. I have also cubes of brass, lead, and ivory; by

counterpoising them with the cubes of water, we can easily find their specific gravities; they are shown together with that of cast iron in the following table:—

Substance.	Specific Gravity.
Cast Iron	7·2
Brass	8·1
Lead	11·3
Ivory	1·8

94. The mode here adopted of finding specific gravities is entirely different from the far more accurate methods which are commonly used, but the explanation of the latter involve more difficult principles than those we have been considering. Our method rather offers an explanation of the nature of specific gravity than a good means of determining it, though, as we have seen, it gives a result sufficiently near the truth for many purposes.

THE PLUMMET AND SPIRIT-LEVEL.

95. The tendency of the earth to draw all bodies towards it is well illustrated by the useful "line and plummet." This consists merely of a string to one end of which a leaden weight is attached. The string when at rest hangs vertically; if the weight be drawn to one side, it will, when released, swing backwards and forwards, until it finally settles again in the vertical; the reason is that the weight always tries to get as near the earth as it can, and this is accomplished when the string hangs vertically downwards.

96. The surface of water in equilibrium is a horizontal plane; that is also a consequence of gravity. All the particles of water try to get as near the earth as possible, and therefore if any portion of the water were higher than the rest, it would immediately spread, as by doing so it could get lower.

IV.] THE CENTRE OF GRAVITY. 57

97. Hence the surface of a fluid at rest enables us to find a perfectly horizontal plane, while the plummet gives us a perfectly vertical line: both these consequences of gravity are of the utmost practical importance.

98. The spirit-level is another common and very useful instrument which depends on gravity. It consists of a glass tube slightly curved, with its convex surface upwards, and attached to a stand with a flat base. This tube is nearly filled with spirit, but a bubble of air is allowed to remain. The tube is permanently adjusted so that, when the plate is laid on a perfectly horizontal surface, the bubble will stand in the middle: accordingly the position of the bubble gives a means of ascertaining whether a surface is level.

THE CENTRE OF GRAVITY.

99. We proceed to an experiment which will give an insight into a curious property of gravity. I have here a plate of sheet iron; it has the irregular shape shown in Fig 27. Five small holes A B C D E are punched at different positions on the margin. Attached to the framework is a small pin from which I can suspend the iron plate by one of its holes A: the plate is not supported in any other way; it hangs freely from the pin, around which it can be easily turned. I find that there is one position, and one only, in which the plate will rest; if I withdraw it from that position it returns there after a few oscillations. In order to mark this position, I suspend a line and plummet from the pin,

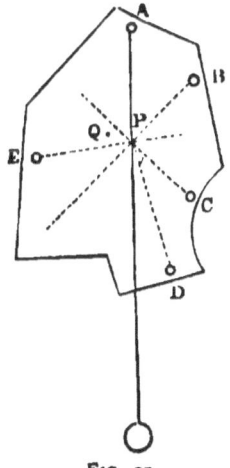

FIG. 27.

having rubbed the line with chalk. I allow the line to come to rest in front of the plate. I then flip the string against the plate, and thus produce a chalked mark: this of course traces out a vertical line A P on the plate.

I now remove the plummet and suspend the plate from another of its holes B, and repeat the process, thus drawing a second chalked line B P across the plate, and so on with the other holes: I thus obtain five lines across the plate, represented by dotted lines in the figure. It is a very remarkable circumstance that these five lines all intersect in the same point P; and if additional holes were bored in the plate, whether in the margin or not, and the chalk line drawn from each of them in the manner described, they would one and all pass through the same point. This remarkable point is called *the centre of gravity* of the plate, and the result at which we have arrived may be expressed by saying that the vertical line from the point of suspension always passes through the centre of gravity.

100. At the centre of gravity P a hole has been bored, and when I place the supporting pin through this hole you see that the plate will rest indifferently in all positions: this is a curious property of the centre of gravity. The centre of gravity may in this respect be contrasted with another hole Q, which is only an inch distant: when I support the plate by this hole, it has only one position of rest, viz. when the centre of gravity P is vertically beneath Q. Thus the centre of gravity differs remarkably from any other point in the plate.

101. We may conceive the force of gravity on the plate to act as a force applied at P. It will then be easily seen why this point remains vertically underneath the point of suspension when the body is at rest. If I attached a string to the plate and pulled it, the plate would evidently place itself so

IV.] STABLE AND UNSTABLE EQUILIBRIUM. 59

that the direction of the string would pass through the point of suspension; in like manner gravity so places the plate that the direction of its force passes through the point of suspension.

102. Whatever be the form of the plate it always contains one point possessing these remarkable properties, and we may state in general that in every body, no matter what be its shape, there is a point called the centre of gravity, such that if the body be suspended from this point it will remain in equilibrium indifferently in any position, and that if the body be suspended from any other point, then it will be in equilibrium when the centre of gravity is directly underneath the point of suspension. In general, it will be impossible to support a body exactly at its centre of gravity, as this point is within the mass of the body, and it may also sometimes happen that the centre of gravity does not lie in the substance of the body at all, as for example in a ring, in which case the centre of gravity is at the centre of the ring. We need not, however, dwell on these exceptional cases, as sufficient illustrations of the truth of the laws mentioned will present themselves subsequently.

STABLE AND UNSTABLE EQUILIBRIUM.

103. An iron rod A B, capable of revolving round an axis passing through its centre P, is shown in Fig. 28.

The centre of gravity lies at the centre B, and consequently, as is easily seen, the rod will remain at rest in whatever position it be placed. But let a weight R be attached to the rod by means of a binding screw. The centre of gravity of the whole is no longer at the centre of the rod; it has moved to a point s nearer the weight; we may easily ascertain its position by removing the rod from its axle and then ascertaining the point about which it will balance. This may be done by placing the bar on a knife-edge, and moving it to and fro until the right position be secured; mark this position on the rod, and return it to its axle, the weight being still attached. We do not now find that the rod will balance in every position. You see it will balance if the point s be directly underneath the axis, but not if it lie to one side or the other. But if s be directly over the axis, as in the figure, the rod is in a curious condition. It will, when carefully placed, remain at rest; but if it receive the slightest displacement, it will tumble over. The rod is in equilibrium in this position, but it is what is called *unstable* equilibrium. If the centre of gravity be vertically below the point of suspension, the rod will return again if moved away: this position is therefore called one of *stable* equilibrium. It is very important to notice the distinction between these two kinds of equilibrium.

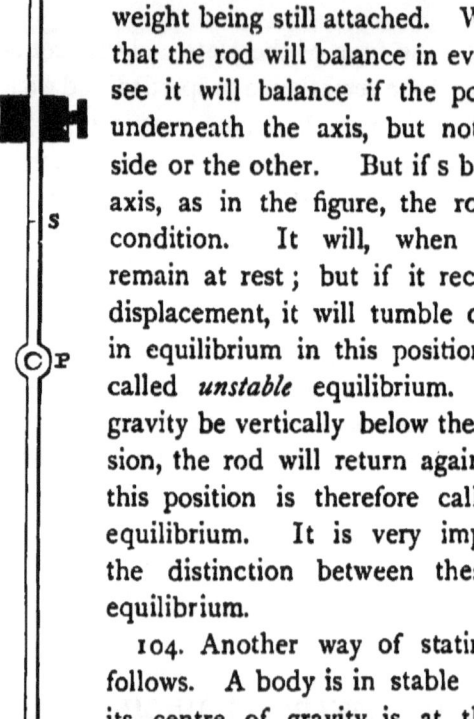

Fig. 28.

104. Another way of stating the case is as follows. A body is in stable equilibrium when its centre of gravity is at the lowest point: unstable when it is at the highest. This may be very simply illustrated by an ellipse, which I hold in my hand. The centre of gravity of this figure is at its centre. The ellipse, when resting on its side, is in a position of stable equilibrium and its centre of gravity is then clearly at its lowest point. But I can also balance the ellipse on its narrow end, though if I do so the smallest touch suffices to overturn it. The ellipse is then in unstable equilibrium; in this case, obviously, the centre of gravity is at the highest point.

CENTRE OF GRAVITY.

105. I have here a sphere, the centre of gravity of which is at its centre; in whatever way the sphere is placed on a plane, its centre is at the same height, and therefore cannot be said to have any highest or lowest point; in such a case as this the equilibrium is *neutral*. If the body be displaced, it will not return to its old position, as it would have done had that been a position of stable equilibrium, nor will it deviate further therefrom as if the equilibrium had been unstable: it will simply remain in the new position to which it is brought.

106. I try to balance an iron ring upon the end of a stick H, Fig. 29, but I cannot easily succeed in doing so. This is because its centre of gravity s is above the point of support; but if I place the stick at F, the ring is in stable equilibrium, for now the centre of gravity is below the point of support.

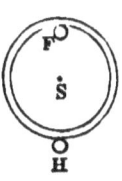

FIG. 29.

PROPERTY OF THE CENTRE OF GRAVITY IN A REVOLVING WHEEL.

107. There are other curious consequences which follow from the properties of the centre of gravity, and we shall conclude by illustrating one of the most remarkable, which is at the same time of the utmost importance in machinery.

108. It is generally necessary that a machine should work as steadily as possible, and that undue vibration and shaking of the framework should be avoided: this is particularly the case when any parts of the machine rotate with great velocity, as, if these be heavy, inconvenient vibration will be produced when the proper adjustments are not made. The connection between this and the centre of

gravity will be understood by reference to the apparatus represented in the accompanying figure (Fig. 30). We have here an arrangement consisting of a large cog wheel C working into a small one B, whereby, when the handle H is turned, a velocity of rotation can be given to the iron

FIG. 30.

disk D, which weighs 14 lbs., and is 18″ in diameter. This disk being uniform, and being attached to the axis at its centre, it follows that its centre of gravity is also the centre of rotation. The wheels are attached to a stand, which, though massive, is still unconnected with the floor. By turning the handle I can rotate the disk very rapidly, even

IV.]   CENTRE OF GRAVITY.   63

as much as twelve times in a second. Still the stand remains quite steady, and even the shutter bell attached to it at E is silent.

109. Through one of the holes in the disk D I fasten a small iron bolt and a few washers, altogether weighing about 1 lb.; that is, only one-fourteenth of the weight of the disk. When I turn the handle slowly, the machine works as smoothly as before; but as I increase the speed up to one revolution every two seconds, the bell begins to ring violently, and when I increase it still more, the stand quite shakes about on the floor. What is the reason of this? By adding the bolt, I slightly altered the position of the centre of gravity of the disk, but I made no change of the axis about which the disk rotated, and consequently the disk was not on this occasion turning round its centre of gravity: this it was which caused the vibration. It is absolutely necessary that the centre of gravity of any heavy piece, rotating rapidly about an axis, should lie in the axis of rotation. The amount of vibration produced by a high velocity may be very considerable, even when a very small mass is the originating cause.

110. In order that the machine may work smoothly again, it is not necessary to remove the bolt from the hole. If by any means I bring back the centre of gravity to the axis, the same end will be attained. This is very simply effected by placing a second bolt of the same size at the opposite side of the disk, the two being at equal distances from the axis; on turning the handle, the machine is seen to work as smoothly as it did in the first instance.

111. The most common rotating pieces in machines are wheels of various kinds, and in these the centre of gravity is evidently identical with the centre of rotation; but if from any cause a wheel, which is to turn rapidly, has an

extra weight attached to one part, this weight must be counterpoised by one or more on other portions of the wheel, in order to keep the centre of gravity of the whole in its proper place. Thus it is that the driving wheels of a locomotive are always weighted so as to counteract the effect of the crank and restore the centre of gravity to the axis of rotation. The cause of the vibration will be understood after the lecture on centrifugal force (Lect. XVII.).

LECTURE V.

THE FORCE OF FRICTION.

The Nature of Friction.—The Mode of Experimenting.—Friction is proportional to the pressure.—A more accurate form of the Law.—The Coefficient varies with the weights used.—The Angle of Friction.—Another Law of Friction.—Concluding Remarks.

THE NATURE OF FRICTION.

112. A DISCUSSION of the force of friction is a necessary preliminary to the study of the mechanical powers which we shall presently commence. Friction renders the inquiry into the mechanical powers more difficult than it would be if this force were absent; but its effects are too important to be overlooked.

113. The nature of friction may be understood by Fig. 31, which represents a section of the top of a table of wood

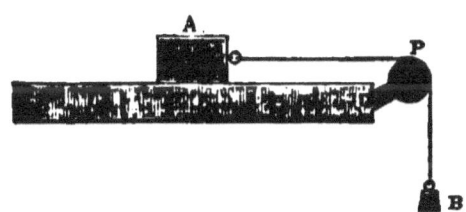

FIG. 31.

or any other substance levelled so that C D is horizontal; on the table rests a block A of wood or any other substance. To A a cord is attached, which, after passing over a pulley P,

F

is stretched by a suspended weight B. If the magnitude of B exceeds a certain limit, then A is pulled along the table and B descends; but if B be smaller than this limit, both A and B remain at rest. When B is not heavy enough to produce motion it is supported by the tension of the cord, which is itself neutralized by the *friction* produced by a certain coherence between A and the table. Friction is by this experiment proved to be a force, because it prevents the motion of B. Indeed friction is generally manifested as a force by destroying motion, though sometimes indirectly producing it.

114. The true source of the force lies in the inevitable *roughness* of all known surfaces, no matter how they may have been wrought. The minute asperities on one surface are detained in corresponding hollows in the other, and consequently force must be exerted to make one surface slide upon the other. By care in polishing the surfaces the amount of friction may be diminished; but it can only be decreased to a certain limit, beyond which no amount of polishing seems to produce much difference.

115. The law of friction under different conditions must be inquired into, in order that we may make allowance when its effect is of importance. The discussion of the experiments is sometimes a little difficult, and the truths arrived at are principally numerical, but we shall find that some interesting laws of nature will appear.

THE MODE OF EXPERIMENTING.

116. Friction is present between every pair of surfaces which are in contact: there is friction between two pieces of wood, and between a piece of wood and a piece of iron; and the amount of the force depends upon the character of both surfaces. We shall only experiment upon the friction

of wood upon wood, as more will be learned by a careful study of a special case than by a less minute examination of a number of pairs of different substances.

117. The apparatus used is shown in Fig. 32. A plank of pine 6' × 11" × 2" is planed on its upper surface, levelled by a spirit-level, and firmly secured to the framework at a height of about 4' from the ground. On it is a pine slide 9" × 9", the grain of which is crosswise to that of the plank; upon the slide the load A is placed. A rope is attached to the slide, which passes over a very freely mounted cast iron pulley C, 14" diameter, and carries at the other end a hook weighing one pound, from which weights B can be suspended.

118. The mode of experimenting consists in placing a certain load upon A, and then ascertaining what weight applied to B will draw the loaded slide along the plane. As several trials are generally necessary to determine the power, a rope is attached to the back of the slide, and passes over the two pulleys D; this makes it easy for the experimenter, when applying the weights at B, to draw back the slide to the end of the plane by pulling the ring E: this rope is of course left quite slack during the process of the experiment, since the slide must not be retarded. The loads placed upon A during the series of experiments ranged between one stone and eight stone. In the loads stated the weight of the slide itself, which was less than 1 lb., is always included. A variety of small weights were provided for the hook B; they consisted of 0·1, 0·5, 1, 2, 7, and 14 lbs. There is some friction to be overcome in the pulley C, but as the pulley is comparatively large its friction is small, though it was always allowed for.

119. An example of the experiments made is thus described. A weight of 56 lbs. is placed upon the slide, and it is found

68   EXPERIMENTAL MECHANICS.   [LECT.

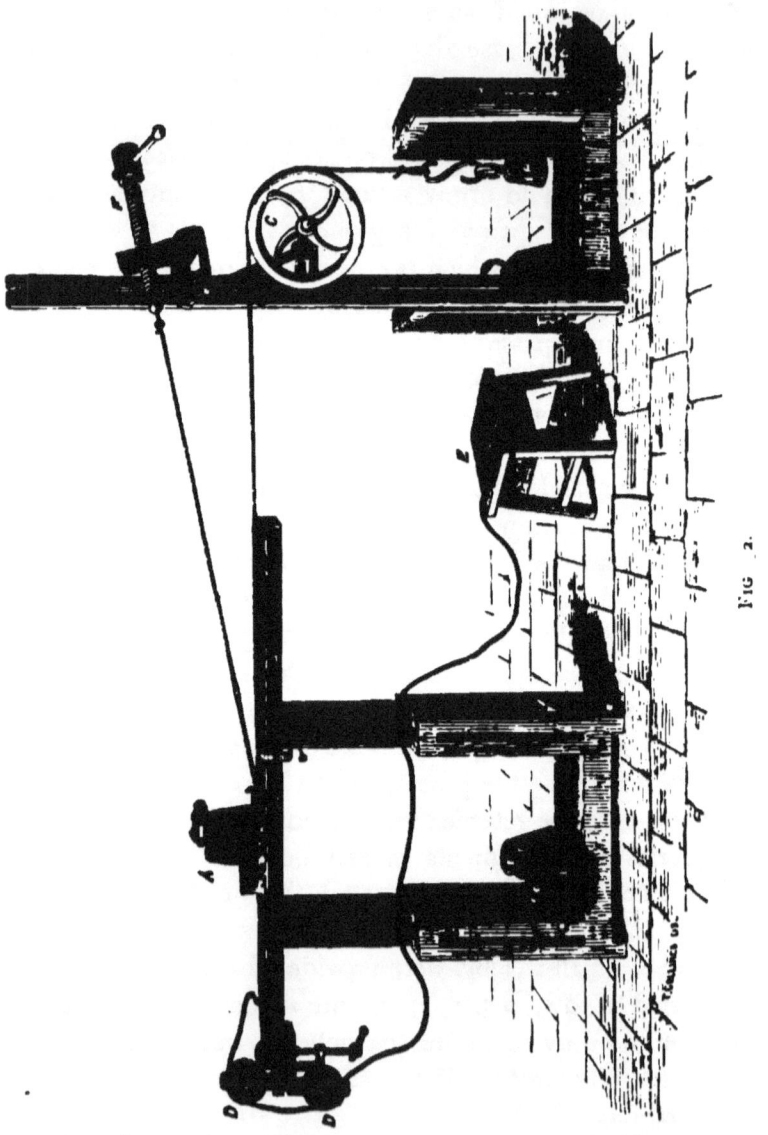

Fig. 2.

on trial that 29 lbs. on B (including the weight of the hook itself) is sufficient to start the slide; this weight is placed upon the hook pound by pound, care being taken to make each addition gently.

120. Experiments were made in this way with various weights upon A, and the results are recorded in Table I.

TABLE I.—FRICTION.

Smooth horizontal surface of pine 72" × 11"; slide also of pine 9" × 9"; grain crosswise; slide is not started; force acting on slide is gradually increased until motion commences.

Number of Experiment.	Load on slide in lbs., including weight of slide.	Force necessary to move slide. 1st Series.	Force necessary to move slide. 2nd Series.	Mean values.
1	14	5	8	6·5
2	28	15	16	15·5
3	42	20	15	17·5
4	56	29	24	26·5
5	70	33	31	32·0
6	84	43	33	35·0
7	98	42	38	40·0
8	112	50	33	41·5

In the first column a number is given to each experiment for convenience of reference. In the second column the load on the slide is stated in lbs. In the third column is found the force necessary to overcome the friction. The fourth column records a second series of experiments performed in the same manner as the first series; while the last column shows the mean of the two frictions.

121. The first remark to be made upon this table is, that the results do not appear satisfactory or concordant. Thus from 6 and 7 of the 1st series it would appear that the friction of 84 lbs. was 43 lbs., while that of 98 lbs. was 42 lbs., so that here the greater weight appears to have the less friction, which is evidently contrary to the whole tenor of the results, as a glance will show. Moreover the frictions in the 1st and the 2nd series do not agree, being generally greater

in the former than in the latter, the discordance being especially noticeable in experiment 8, where the results were 50 lbs. and 33 lbs. In the final column of means these irregularities are lessened, for this column shows that the friction increases with the weight, but it is sufficient to observe that as the difference of the 1st and the 2nd is 9 lbs., and that of the 2nd and the 3rd is only 2 lbs., the discovery of any law from these results is hopeless.

122. But is friction so capricious that it is amenable to no better law than these experiments appear to indicate? We must look a little more closely into the matter. When two pieces of wood have remained in contact and at rest for some time, a second force besides friction resists their separation : the wood is compressible, the surfaces become closely approximated, and the coherence due to this cause must be overcome before motion commences. The initial coherence is uncertain ; it depends probably on a multitude of minute circumstances which it is impossible to estimate, and its presence has vitiated the results which we have found so unsatisfactory.

123. We can remove these irregularites by *starting* the slide at the commencement. This may be conveniently effected by the screw shown at F in Fig. 32; a string attached to its end is fastened to the slide, and by giving the handle of the screw a few turns the slide begins to creep. A body once set in motion will continue to move with the same velocity unless acted upon by a force; hence the weight at B just overcomes the friction when the slide moves uniformly after receiving a start: this velocity was in one case of average speed measured to be 16 inches per minute.

124. Indeed in no case can the slide *commence* to move unless the force *exceed* the friction. The amount of this

V.] THE MODE OF EXPERIMENTING. 71

excess is indeterminate. It is certainly greater between wooden surfaces than between less compressible surfaces like those of metals or glass. In the latter case, when the force exceeds the friction by a small amount, the slide starts off with an excessively slow motion; with wood the force must exceed the friction by a larger amount before the slide commences to move, but the motion is then comparatively rapid.

125. If the power be too small, the load either does not continue moving after the start, or it stops irregularly. If the power be too great, the load is drawn with an accelerated velocity. The correct amount is easily recognised by the uniformity of the movement, and even when the slide is heavily laden, a few tenths of a pound on the power hook cause an appreciable difference of velocity.

126. The accuracy with which the friction can be measured may be appreciated by inspecting Table II.

TABLE II.—FRICTION.

Smooth horizontal surface of pine 72″ × 11″; slide also of pine 9″ × 9″; grain crosswise; slide started; force applied is sufficient to maintain uniform motion of the slide.

Number of Experiment.	Load on slide in lbs., including weight of slide.	Force necessary to maintain motion. 1st Series.	Force necessary to maintain motion. 2nd Series.	Mean values.
1	14	4·9	4·9	4·9
2	28	8·5	8·6	8·5
3	42	12·6	12·4	12·5
4	56	16·3	16·2	16·2
5	70	19·7	20·0	19·8
6	84	23·7	23·0	23·4
7	98	26·5	26·1	26·3
8	112	29·7	29·9	29·8

127. Two series of experiments to determine the power necessary to maintain the motion have been recorded.

Thus, in experiment 7, the load on the slide being 98 lbs., it was found that 26·5 lbs. was sufficient to sustain the motion, and a second trial being made independently, the power found was 26·1 lbs.: a mean of the two values, 26·3 lbs., is adopted as being near the truth. The greatest difference between the two series, amounting to 0·7 lb., is found in experiment 6; a third value was therefore obtained for the friction of 84 lbs.: this amounted to 23·5 lbs., which is intermediate between the two former results, and 23·4 lbs., a mean of the three, is adopted as the final result.

128. The close accordance of the experiments in this table shows that the means of the fifth column are probably very near the true values of the friction for the corresponding loads upon the slide.

129. The mean frictions must, however, be slightly diminished before we can assert that they represent only the friction of the wood upon the wood. The pulley over which the rope passes turns round its axle with a small amount of friction, which must also be overcome by the power. The mode of estimating this amount, which in these experiments never exceeds 0·5 lb., need not now be discussed. The corrected values of the friction are shown in the third column of Table III. Thus, for example, the 4·9 lbs. of friction in experiment 1 consists of 4·7, the true friction of the wood, and 0·2, which is the friction of the pulley; and 26·3 of experiment 7 is similarly composed of 25·8 and 0·5. It is the corrected frictions which will be employed in our subsequent calculations.

FRICTION IS PROPORTIONAL TO THE PRESSURE.

130. Having ascertained the values of the force of friction for eight different weights, we proceed to inquire into the

laws which may be founded on our results. It is evident that the friction increases with the load, of which it is always greater than a fourth, and less than a third. It is natural to surmise that the friction (F) is really a constant fraction of -the load (R)—in other words, that $F = kR$, where k is a constant number.

131. To test this supposition we must try to determine k; it may be ascertained by dividing any value of F by the corresponding value of R. If this be done, we shall find that each of the experiments yields a different quotient; the first gives 0·336, and the last 0·262, while the other experiments give results between these extreme values. These numbers are tolerably close together, but there is still sufficient discrepancy to show that it is not strictly true to assert that the friction is proportional to the load.

132. But the law that the *friction varies proportionally to the pressure* is so approximately true as to be sufficient for most practical purposes, and the question then arises, which of the different values of k shall we adopt? By a method which is described in the Appendix we can determine a value for k which, while it does not represent any one of the experiments precisely, yet represents them collectively better than it is possible for any other value to do. The number thus found is 0·27. It is intermediate between the two values already stated to be extreme. The character of this result is determined by an inspection of Table III.

The fourth column of this table has been calculated from the formula $F = 0·27\ R$. Thus, for example, in experiment 5, the friction of a load of 70 lbs. is 19·4 lbs., and the product of 70 and 0·27 is 18·9, which is 0·5 lb. less than the true amount. In the last column of this table the discrepancies between the observed and the calculated values are recorded, for facility of comparison. It will be observed that the greatest difference is under 1 lb.

TABLE III.—FRICTION.

Friction of pine upon pine; the mean values of the friction given in Table II. (corrected for the friction of the pulley) compared with the formula $F = 0.27 R$.

Number of Experiment.	R. Total load on slide in lbs.	Corrected mean value of friction.	F. Calculated value of friction.	Discrepancies between the observed and calculated frictions
1	14	4.7	3.8	−0.9
2	28	8.2	7.6	−0.6
3	42	12.2	11.3	−0.9
4	56	15.8	15.1	−0.7
5	70	19.4	18.9	−0.5
6	84	23.0	22.7	−0.3
7	98	25.8	26.5	+0.7
8	112	29.3	30.2	+0.9

133. Hence the law $F = 0.27 R$ represents the experiments with tolerable accuracy; and the numerical ratio 0.27 is called the *coefficient of friction*. We may apply this law to ascertain the friction in any case where the load lies between 14 lbs. and 112 lbs.; for example, if the load be 63 lbs., the friction is 63 × 0.27 = 17.0.

134. The coefficient of friction would have been slightly different had the grain of the slide been parallel to that of the plank; and it of course varies with the nature of the surfaces. Experimenters have given tables of the coefficients of friction of various substances, wood, stone, metals, &c. The use of these coefficients depends upon the assumption of the ordinary law of friction, namely, that the friction is proportional to the pressure: this law is accurate enough for most purposes, especially when used for loads that lie between the extreme weights employed in calculating the value of the coefficient which is employed.

A MORE ACCURATE LAW OF FRICTION.

135. In making one of our measurements with care, it is unusual to have an error of more than a few tenths of 1 lb. and it is hardly possible that any of the *mean frictions* we have found should be in error to so great an extent as 0·5 lb. But with the value of the coefficient of friction which is used in Table III., the discrepancies amount sometimes to 0·9 lbs. With any other numerical coefficient than 0·27, the discrepancies would have been even still more serious. As these are too great to be attributed to errors of experiment, we have to infer that the law of friction which has been assumed cannot be strictly true. The signs of the discrepancies indicate that the law gives frictions which for small loads are too small, and for large loads are too great.

136. We are therefore led to inquire whether some other relation between F and R may not represent the experiments with greater fidelity than the common law of friction. If we diminished the coefficient by a small amount, and then added a constant quantity to the product of the coefficient and the load, the effect of this change would be that for small loads the calculated values would be increased, while for large loads they would be diminished. This is the kind of change-which we have indicated to be necessary for reconciliation between the observed and calculated values.

137. We therefore infer that a relation of the form $F = x + yR$ will probably express a more correct law, provided we can find x and y. One equation between x and y is obtained by introducing any value of R with the corresponding value of F, and a second equation can be found by taking any other similar pair. From these two equations the values of x and of y may be deduced by elementary algebra, but the best formula will be obtained

by combining together all the pairs of corresponding values. For this reason the method described in the Appendix must be used, which, as it is founded on all the experiments, must give a thoroughly representative result. The formula thus determined, is

$$F = 1\cdot44 + 0\cdot252\ R.$$

This formula is compared with the experiments in Table IV.

TABLE IV.—FRICTION.

Friction of pine upon pine; the mean values of the friction given in Table II. (corrected for the friction of the pulley) compared with the formula $F = 1\cdot44 + 0\cdot252\ R$.

Number of Experiment.	R. Total load on slide in lbs.	Corrected mean value of friction.	F. Calculated value of friction.	Discrepancies between the observed and calculated frictions
1	14	4·7	5·0	+0·3
2	28	8·2	8·5	+0·3
3	42	12·2	12·0	−0·2
4	56	15·8	15·6	−0·2
5	70	19·4	19·1	−0·3
6	84	23·0	22·6	−0·4
7	98	25·8	26·1	+0·3
8	112	29·3	29·7	+0·4

The fourth column contains the calculated values: thus, for example, in experiment 4, where the load is 56 lbs., the calculated value is $1\cdot44 + 0\cdot252 \times 56 = 15\cdot6$; the difference 0·2 between this and the observed value 15·8 is shown in the last column.

138. It will be noticed that the greatest discrepancy in this column is 0·4 lbs., and that therefore the formula represents the experiments with considerable accuracy. It is undoubtedly nearer the truth than the former law (Art. 132); in fact, the differences are now such as might really belong to errors unavoidable in making the experiments.

v.] COEFFICIENT VARIES WITH WEIGHTS. 77

139. This formula may be used for calculating the friction for any load between 14 lbs. and 112 lbs. Thus, if the load be 63 lbs., the friction is $1·44 + 0·252 \times 63 = 17·3$ lbs., which does not differ much from 17·0 lbs., the value found by the more ordinary law. We must, however, be cautious not to apply this formula to weights which do not lie between the limits of the greatest and least weight used in those experiments by which the numerical values in the formula have been determined; for example, to take an extreme case, if $R = 0$, the formula would indicate that the friction was 1·44, which is evidently absurd; here the formula errs in excess, while if the load were very large it is certain the formula would err in defect.

THE COEFFICIENT VARIES WITH THE WEIGHTS USED.

140. In a subsequent lecture we shall employ as an inclined plane the plank we have been examining, and we shall require to use the knowledge of its friction which we are now acquiring. The weights which we shall then employ range from 7 lbs. to 56 lbs. Assuming the ordinary law of friction, we have found that 0·27 is the best value of its coefficient when the loads range between 14 lbs. and 112 lbs. Suppose we only consider loads up to 56 lbs., we find that the coefficient 0·288 will best represent the experiments within this range, though for 112 lbs. it would give an error of nearly 3 lbs. The results calculated by the formula $F = 0·288\,R$ are shown in Table V., where the greatest difference is 0·7 lb.

141. But we can replace the common law of friction by the more accurate law of Art. 137, and the formula computed so as to best harmonise the experiments up to 56 lbs., disregarding the higher loads, is $F = 0·9 + 0·266\,R$. This

TABLE V.—FRICTION.

Friction of pine upon pine; the mean values of the friction given in Table II. (corrected for the friction of the pulley) compared with the formula $F = 0\cdot288\,R$

Number of Experiment.	R. Total load on slide in lbs.	Corrected mean value of friction.	F. Calculated value of friction.	Discrepancies between the observed and calculated frictions
1	14	4·7	4·0	−0·7
2	28	8·2	8·1	−0·1
3	42	12·2	12·1	−0·1
4	56	15·8	16·1	+0·3

formula is obtained by the method referred to in Art. 137. We find that it represents the experiments better than that used in Table V. Between the limits named, this formula is also more accurate than that of Table IV. It is compared with the experiments in Table VI., and it will be noticed that it represents them with great precision, as no discrepancy exceeds 0·1.

TABLE VI.—FRICTION.

Friction of pine upon pine; the mean values of the friction given in Table II. (corrected for the friction of the pulley) compared with the formula $F = 0\cdot9 + 0\cdot266\,R$.

Number of Experiment.	R. Total Load on slide in lbs.	Corrected mean value of friction.	F. Calculated value of friction.	Discrepancies between the observed and calculated frictions
1	14	4·7	4·6	−0·1
2	28	8·2	8·3	+0·1
3	42	12·2	12·1	−0·1
4	56	15·8	15·8	0·0

THE ANGLE OF FRICTION.

142. There is another mode of examining the action of friction besides that we have been considering. The apparatus for this purpose is shown in Fig. 33, in which B C represents the plank of pine we have already used. It is

v.] THE ANGLE OF FRICTION. 79

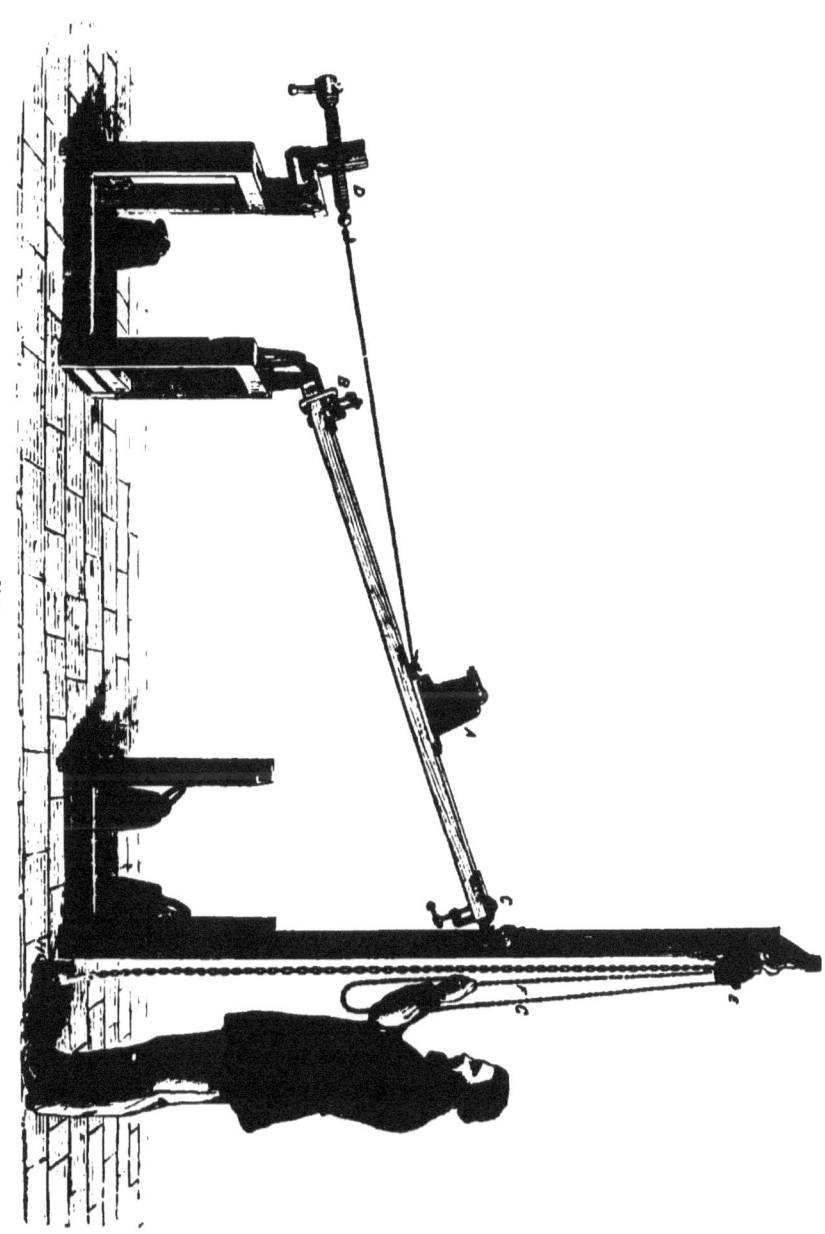

FIG. 33.

now mounted so as to be capable of turning about one end B; the end C is suspended from one hook of the chain from the "*epicycloidal*" pulley-block E. This block is very convenient for the purpose. By its means the inclination of the plank can be adjusted with the greatest nicety, as the raising chain G is held in one hand and the lowering chain F in the other. Another great convenience of this block is that the load does not run down when the lifting chain is left free. The plank is clamped to the hinge about which it turns. The frames by which both the hinge and the block are supported are weighted in order to secure steadiness. The inclination of the plane is easily ascertained by measuring the difference in height of its two ends above the floor, and then making a drawing on the proper scale. The starting-screw D, whose use has been already mentioned, is also fastened to the frame-work in the position shown in the figure.

143. Suppose the slide A be weighted and placed upon the inclined plane B C; if the end C be only slightly elevated, the slide remains at rest; the reason being that the friction between the slide and the plane neutralizes the force of gravity. But suppose, by means of the pulley-block, C be gradually raised; an elevation is at last reached at which the slide starts off, and runs with an accelerating velocity to the bottom of the plane. The angle of elevation of the plane when this occurs is called the angle of statical friction.

144. The weights with which the slide was laden in these experiments were 14 lbs., 56 lbs., and 112 lbs., and the results are given in Table VII.

We see that a load of 56 lbs. started when the plane reached an inclination of 20°·1 in the first series, and of 17°·2 in the second, the mean value 18°·6 being given in the fifth column. These means for the three different weights

THE ANGLE OF FRICTION.

agree so closely that we assert the remarkable law that *the angle of friction does not depend upon the magnitude of the load.*

TABLE VII.—ANGLE OF STATICAL FRICTION.

A smooth plane of pine 72″ × 11″ carries a loaded slide of pine 9″ × 9″; one end of the plane is gradually elevated until the slide starts off.

Number of Experiment.	Total load on the slide in lbs.	Angle of elevation. 1st Series.	Angle of elevation. 2nd Series.	Mean values of the angles.
1	14	19°·5	—	19°·5
2	56	20°·1	17°·2	18°·6
3	112	20°·3	18°·9	19°·6

145. We might, however, proceed differently in determining the angle of friction, by giving the load a start, and ascertaining if the motion will continue. To do so requires the aid of an assistant, who will start the load with the help of the screw, while the elevation of the plane is being slowly increased. The result of these experiments is given in Table VIII.

TABLE VIII.—ANGLE OF FRICTION.

A smooth plane of pine 72″ × 11″ carries a loaded slide of pine 9″ × 9″; one end of the plane is gradually elevated until the slide, having received a start, moves off uniformly.

Number of Experiment.	Total load on the slide in lbs.	Angle of inclination.
1	14	14°·3
2	56	13°·0
3	112	13°·0

We see from this table also that the angle of friction is independent of the load, but the angle is in this case less by 5° or 6° than was found necessary to impart motion when a start was not given.

G

146. It is commonly stated that the coefficient of friction equals the tangent of the angle of friction, and this can be proved to be true when the ordinary law of friction is assumed. But as we have seen that the law of friction is only approximately correct, we need not expect to find this other law completely verified.

147. When the slide is started, the mean value of the angle of friction is $13°·4$. The tangent of this angle is $0·24$: this is about 11 per cent. less than the coefficient of friction $0·27$, which we have already determined. The mean value of the angle of friction when the slide is not started is $19°·2$, and its tangent is $0·35$. The experiments of Table I. are, as already pointed out, rather unsatisfactory, but we refer to them here to show that, so far as they go, the coefficient of friction is in no sense equal to the tangent of the angle of friction. If we adopt the mean values given in the last column of Table I., the best coefficient of friction which can be deduced is $0·41$. Whether, therefore, the slide be started or not started, the tangent of the angle of friction is smaller than the corresponding coefficient of friction. When the slide is started, the tangent is about 11 per cent. less than the coefficient; and when the slide is not started, it is about 14 per cent. less. There are doubtless many cases in which these differences are sufficiently small to be neglected, and in which, therefore, the law may be received as true.

ANOTHER LAW OF FRICTION.

148. The area of the wooden slide is $9'' \times 9''$, but we would have found that the friction under a given load was practically the same whatever were the area of the slide, so long as its material remained unaltered. This follows as a consequence of

the approximate law that the friction is proportional to the pressure. Suppose that the weight were 100 lbs., and the area of the slide 100 inches, there would then be a pressure cf 1 lb. per square inch over the surface of the slide, and therefore the friction to be overcome on each square inch would be 0·27 lb., or for the whole slide 27 lbs. If, however, the slide had only an area of 50 square inches, the load would produce a pressure of 2 lbs., per square inch; the friction would therefore be 2 × 0·27 = 0·54 lb. for each square inch, and the total friction would be 50 × 0·54 = 27 lbs., the same as before: hence the total friction is independent of the extent of surface. This would remain equally true even though the weight were not, as we have supposed, uniformly distributed over the surface of the slide.

CONCLUDING REMARKS.

149. The importance of friction in mechanics arises from its universal presence. We often recognize it as a destroyer or impeder of motion, as a waster of our energy, and as a source of loss or inconvenience. But, on the other hand, friction is often indirectly the means of producing motion, and of this we have a splendid example in the locomotive engine. - The engine being very heavy, the wheels are pressed closely to the rails; there is friction enough to prevent the wheels slipping, consequently when the engines force the wheels to turn round they must roll onwards. The coefficient of friction of wrought iron upon wrought iron is about 0·2. Suppose a locomotive weigh 30 tons, and the share of this weight borne by the driving wheels be 10 tons, the friction between the driving wheels and the rails is 2 tons. This is the greatest force the engine can exert on a level line. A force of 10 lbs. for every ton

weight of the train is known to be sufficient to sustain the motion, consequently the engine we have supposed should draw along the level a load of 448 tons.

150. But we need not invoke the steam engine to show the use of friction. We could not exist without it. In the first place we could not move about, for walking is only possible on account of the friction between the soles of our boots and the ground; nor if we were once in motion could we stop without coming into collision with some other object, or grasping something to hold on by. Objects could only be handled with difficulty, nails would not remain in wood, and screws would be equally useless. Buildings could hardly be erected, nay, even hills and mountains would gradually disappear, and finally dry land would be immersed beneath the level of the sea. Friction is, so far as we are concerned, quite as essential a law of nature as the law of gravitation. We must not seek to evade it in our mechanical discussions because it makes them a little more difficult. Friction obeys laws; its action is not vague or uncertain. When inconvenient it can be diminished, when useful it can be increased; and in our lectures on the mechanical powers, to which we now proceed, we shall have opportunities of describing machines which have been devised in obedience to its laws.

LECTURE VI.

THE PULLEY.

Introduction.—Friction between a Rope and an Iron Bar.—The use of the Pulley.—Large and Small Pulleys.—The Law of Friction in the Pulley.—Wheels.—Energy.

INTRODUCTION.

151. THE pulley forms a good introduction to the important subject of the mechanical powers. But before entering on the discussions of the next few chapters, it will be necessary for us to explain what is meant in mechanics by "work," and by "energy," which is the capacity for performing work, and we shall therefore include a short outline of this subject in the present lecture.

152. The pulley is a machine which is employed for the purpose of changing the direction of a force. We frequently wish to apply a force in a different direction from that in which it is convenient to exert it, and the pulley enables us to do so. We are not now speaking of these arrangements for increasing power in which pulleys play an important part; these will be considered in the next lecture: we at present refer only to change of direction. In fact, as we shall shortly

see, some force is even wasted when the single fixed pulley is used, so that this machine certainly cannot be called a mechanical power.

153. The occasions upon which a single fixed pulley is used are numerous and familiar. Let us suppose a sack of corn has to be elevated from the lower to one of the upper stories of a building. It may of course be raised by a man who carries it, but he has to lift his own weight in addition to that of the sack, and therefore the quantity of exertion used is greater than absolutely necessary. But supposing there be a pulley at the top of the building over which a rope passes; then, if a man attach one end of the rope to the sack and pull the other, he raises the sack without raising his own weight. The pulley has thus provided the means by which the downward force has been changed in direction to an upward force.

154. The weights, ropes, and pulleys which are used in our windows for counterpoising the weight of the sash afford a very familiar instance of how a pulley changes the direction of a force. Here the downward force of the weight is changed by means of the pulley into an upward force, which nearly counterbalances the weight of the sash.

FRICTION BETWEEN A ROPE AND AN IRON BAR.

155. Every one is familiar with the ordinary form of the pulley; it consists of a wheel capable of turning freely on its axle, and it has a groove in its circumference in which the rope lies. But why is it necessary to give the pulley this form? Why could not the direction of the rope be changed by simply passing it over a bar, as well as by the more complicated pulley? We shall best answer this question by actually trying the experiment, which we can do by means of the apparatus of Fig. 34 (see page 90). In this are shown

two iron studs, G, H, 0"·6 diameter, and about 8" apart; over these passes a rope, which has a hook at each end. If I suspend a weight of 14 lbs. from one hook A, and pull the hook B, I can by exerting sufficient force raise the weight on A, but with this arrangement I am conscious of having to exert a very much larger force than would have been necessary to raise 14 lbs. by merely lifting it.

156. In order to study the question exactly, we shall ascertain what weight suspended from the hook B will suffice to raise A. I find that in order to raise 14 lbs. on A no less than 47 lbs. is necessary on B, consequently there is an enormous loss of force: more than two-thirds of the force which is exerted is expended uselessly. If instead of the 14 lbs. weight I substitute any other weight, I find the same result, viz. that more than three times its amount is necessary to raise it by means of the rope passing over the studs. If a labourer, in raising a sack, were to pass a rope over two bars such as these, then for every stone the sack weighed he would have to exert a force of more than three stones, and there would be a very extravagant loss of power.

157. Whence arises this loss? The rope in moving slides over the surface of the iron studs. Although these are quite smooth and polished, yet when there is a strain on the rope it presses closely upon them, and there is a certain amount of force necessary to make the rope slide along the iron. In other words, when I am trying to raise up 14 lbs. with this contrivance, I not only have its weight opposed to me, but also another force due to the sliding of the rope on the iron: this force is due to friction. Were it not for friction, a force of 14 lbs. on one hook would exactly balance 14 lbs. on the other, and the slightest addition to either weight would make it descend and raise the other. If, then, we are obliged to change the direction of a force, we must devise some means

of doing so which does not require so great a sacrifice as the arrangement we have just used.

THE USE OF THE PULLEY.

158. We shall next inquire how it is that we are enabled to obviate friction by means of a pulley. It is evident we must provide an arrangement in which the rope shall not be required to slide upon an iron surface. This end is attained by the pulley, of which we may take 1, Fig. 34, as an example. This represents a cast iron wheel 14″ in diameter, with a V-shaped groove in its circumference to receive the rope: this wheel turns on a ⅝-inch wrought iron axle, which is well oiled. The rope used is about 0″·25 in diameter.

159. From the hooks E, F at each end of the rope a 14lb. weight is suspended. These equal weights balance each other. According to our former experiment with the studs, it would be necessary for me to treble the weight on one of these hooks in order to raise the other, but now I find that an additional 0·5 lb. placed on either hook causes it to descend and make the other ascend. This is a great improvement; 0·5 lb. now accomplishes what 33 lbs. was before required for. We have avoided a great deal of friction, but we have not got rid of it altogether, for 0·25 lb. is incompetent, when added to either weight, to make that weight descend.

160. To what is the improvement due? When the weight descends the rope does not slide upon the wheel, but it causes the wheel to revolve with it, consequently there is little or no friction at the circumference of the pulley; the friction is transferred to the axle. We still have some resistance to overcome, but for smooth oiled iron axles the friction is very small, hence the advantage of the pulley.

LARGE AND SMALL PULLEYS.

There is in every pulley a small loss of power from the force expended in bending the rope; this need not concern us at present, for with the pliable plaited rope that we have employed the effect is inappreciable, but with large strong ropes the loss becomes of importance. The amount of loss by using different kinds of ropes has been determined by careful experiments.

LARGE AND SMALL PULLEYS.

161. There is often a considerable advantage obtained by using large rather than small pulleys. The amount of force necessary to overcome friction varies inversely as the size of the pulley. We shall demonstrate this by actual experiment with the apparatus of Fig. 34. A small pulley K is attached to the large pulley I; they are in fact one piece, and turn together on the same axle. Hence if we first determine the friction with the rope over the large pulley, and then with the rope over the small pulley, any difference can only be due to the difference in size, as all the other circumstances are the same.

162. In making the experiments we must attend to the following point. The pulleys and the socket on which they are mounted weigh several pounds, and consequently there is friction on the axle arising from the weight of the pulleys, quite independently of any weights that may be placed on the hooks. We must then, if possible, evade the friction of the pulley itself, so that the amount of friction which is observed will be entirely due to the weights raised. This can be easily done. The rope and hooks being on the large pulley I, I find that 0·16 lb. attached to one of the hooks, E, is sufficient to overcome the friction of the pulley, and to make that hook descend and raise F. If therefore we leave 0·16 lb. on E, we may consider the

friction due to the weight of the pulley, rope, and hooks as neutralized.

163. I now place a stone weight on each of the hooks E and F. The amount necessary to make the hook E and its load descend is 0·28 lb. This does not of course

FIG. 34.

include the weight of 0·16 lb. already referred to. We see therefore that with the large pulley the amount of friction to be overcome in raising one stone is 0·28 lb.

164. Let us now perform precisely the same experiment with the small pulley. I transfer the same rope and hooks

to K, and I find that 0·16 lb. is not now sufficient to overcome the friction of the pulley, but I add on weights until C will just descend, which occurs when the load reaches 0·95 lb. This weight is to be left on C as a counterpoise, for the reasons already pointed out. I place a stone weight on C and another on D, and you see that C will descend when it receives an additional load of 1·35 lbs.; this is therefore the amount of friction to be overcome when a stone weight is raised over the pulley K.

165. Let us compare these results with the dimensions of the pulleys. The proper way to measure the effective circumference of a pulley when carrying a certain rope is to measure the length of that rope which will just embrace it. The length measured in this way will of course depend to a certain extent upon the size of the rope. I find that the circumferences of the two pulleys are 43"·0 and 9"·5. The ratio of these is 4·5; the corresponding resistances from friction we have seen to be 0·28 lb. and 1·35 lbs. The larger of these quantities is 4·8 times the smaller. This number is very close to 4·5; we must not, as already explained, expect perfect accuracy in experiments in friction. In the present case the agreement is within the 1-16th of the whole, and we may regard it as a proof of the law that *the friction of a pulley is inversely proportional to its circumference.*

166. It is easy to see the reason why friction should diminish when the size of the pulley is increased. The friction acts at the circumference of the axle about which the wheel turns; it is there present as a force tending to retard motion. Now the larger the wheel the greater will be the distance from the axis at which the force acts which overcomes the friction, and therefore the less need be the magnitude of the force. You will perhaps understand

this better after the principle of the lever has been discussed.

167. We may deduce from these considerations the practical maxim that large pulleys are economical of power. This rule is well known to engineers; large pulleys should be used, not only for diminishing friction, but also to avoid loss of power by excessive bending of the rope. A rope is bent gradually around the circumference of a large pulley with far less force than is necessary to accommodate it to a smaller pulley: the rope also is apt to become injured by excessive bending. In coal pits the trucks laden with coal are hoisted to the surface by means of wire ropes which pass from the pit over a pulley into the engine-house: this pulley is of very large dimensions, for the reasons we have pointed out.

THE LAW OF FRICTION IN THE PULLEY.

168. I have here a wooden pulley $3''\cdot 5$ in diameter; the hole is lined with brass, and the pulley turns very freely on an iron spindle. I place the rope and hooks upon the groove. Brass rubbing on iron has but little friction, and when 7 lbs. is placed on each hook, 0·5 lb. added to either will make it descend and raise up the other. Let 14 lbs. be placed on each hook, 0·5 lb. is no longer sufficient; 1 lb. is required: hence when the weight is doubled the friction is also doubled. Repeating the experiment with 21 lbs. and 28 lbs. on each side, the corresponding weights necessary to overcome friction are 1·5 lbs. and 2 lbs. In the four experiments the weights used are in the proportion 1, 2, 3, 4; and the forces necessary to overcome friction, 0·5 lb., 1 lb., 1·5 lbs., and 2 lbs., are in the same proportion. Hence the friction is proportional to the load.

WHEELS.

169. The wheel is one of the most simple and effective

devices for overcoming friction. A sleigh is an admirable vehicle on a smooth surface such as ice, but it is totally unadapted for use on common roads; the reason being that the amount of friction between the sleigh and the road is so great that to move the sleigh the horse would have to exert a force which would be very great compared with the load he was drawing. But a vehicle properly mounted on wheels moves with the greatest ease along the road, for the circumference of the wheel does not slide, and consequently there is no friction between the wheel and the road; the wheel however turns on its axle, therefore there is sliding, and consequently friction, at the axle, but the axle and the wheel are properly fitted to each other, and the surfaces are lubricated with oil, so that the friction is extremely small.

170. With large wheels the amount of friction on the axle is less than with small wheels; other advantages of large wheels are that they do not sink much into depressions in the roads, and that they have also an increased facility in surmounting the innumerable small obstacles from which even the best road is not free.

171. When it is desired to make a pulley turn with extremely small friction, its axle, instead of revolving in fixed bearings, is mounted upon what are called friction wheels. A set of friction wheels is shown in the apparatus of Fig. 66: as the axle revolves, the friction between the axles and the wheels causes the latter to turn round with a comparatively slow motion; thus all the friction is transferred to the axles of the four friction wheels; these revolve in their bearings with extreme slowness, and consequently the pulley is but little affected by friction. The amount of friction in a pulley so mounted may be understood from the following experiment. A silk cord is placed on the pulley,

and 1 lb. weight is attached to each of its ends: these of course balance. A number of fine wire hooks, each weighing 0·001 lb., are prepared, and it is found that when a weight of 0·004 lb. is attached to either side it is sufficient to overcome friction and set the weights in motion.

ENERGY.

172. In connection with the subject of friction, and also as introductory to the mechanical powers, the notion of "work," or as it is more properly called "energy," is of great importance. The meaning of this word as employed in mechanics will require a little consideration.

173. In ordinary language, whatever a man does that can cause fatigue, whether of body or mind, is called work. In mechanics, we mean by energy that particular kind of work which is directly or indirectly equivalent to raising weights.

174. Suppose a weight is lying on the floor and a stool is standing beside it: if a man raise the weight and place it upon the stool, the exertion that he expends is energy in the sense in which the word is used in mechanics. The amount of exertion necessary to place the weight upon the stool depends upon two things, the magnitude of the weight and the height of the stool. It is clear that both these things must be taken into account, for although we know the weight which is raised, we cannot tell the amount of exertion that will be required until we know the height through which it is to be raised; and if we know the height, we cannot appreciate the quantity of exertion until we know the weight.

175. The following plan has been adopted for expressing quantities of energy. The small amount of exertion necessary to raise 1 lb. avoirdupois through one British

foot is taken as a standard, compared with which all other quantities of energy are estimated. This quantity of exertion is called in mechanics the unit of energy, and sometimes also the "foot-pound."

176. If a weight of 1 lb. has to be raised through a height of 2 feet, or a weight of 2 lbs. through a height of 1 foot, it will be necessary to expend twice as much energy as would have raised a weight of 1 lb. through 1 foot, that is, 2 foot-pounds.

If a weight of 5 lbs. had to be raised from the floor up to a stool 3 feet high, how many units of energy would be required? To raise 5 lbs. through 1 foot requires 5 foot-pounds, and the process must be again repeated twice before the weight arrive at the top of the stool. For the whole operation 15 foot-pounds will therefore be necessary.

If 100 lbs. be raised through 20 feet, 100 foot-pounds of energy is required for the first foot, the same for the second, third, &c., up to the twentieth, making a total of 2,000 foot-pounds.

Here is a practical question for the sake of illustration. Which would it be preferable to hoist, by a rope passing over a single fixed pulley, a trunk weighing 40 lbs. to a height of 20 feet, or a trunk weighing 50 lbs. to a height of 15 feet? We shall find how much energy would be necessary in each case: 40 times 20 is 800; therefore in the first case the energy would be 800 foot-pounds. But 50 times 15 is 750; therefore the amount of work, in the second case, is only 750 lbs. Hence it is less exertion to carry 50 lbs. up 15 feet than 40 lbs. up 20 feet.

177. The rate of working of every source of energy, whether it lie in the muscles of men or other animals, in water-wheels, steam-engines, or other prime movers, is to

be measured by the number of foot-pounds produced in the unit of time.

The power of a steam-engine is defined by its equivalent in horse-power. For example, it is meant that a steam-engine of 3 horse-power, could, when working for an hour, do as much work as 3 horses could do when working for the same time. The power of a horse is, however, an uncertain quantity, differing in different animals and not quite uniform in the same individual; accordingly the selection of this measure for the efficiency of the steam-engine is inconvenient. We replace it by a convenient standard horse-power, which is, however, a good deal larger than that continuously obtainable from any ordinary horse. A one horse-power steam-engine is capable of accomplishing 33,000 foot-pounds per minute.

178. We shall illustrate the numerical calculation of horse-power by an example : if a mine be 1,000 feet deep, how much water per minute would a 50 horse-power engine be capable of raising to the surface? The engine would yield 50 × 33,000 units of work per minute, but the weight has to be raised 1,000 feet, consequently the number of pounds of water raised per minute is

$$\frac{50 \times 33,000}{1,000} = 1,650.$$

179. We shall apply the principle of work to the consideration of the pulley already described (p. 90). In order to raise the weight of 14 lbs., it is necessary that the rope to which the power is applied should be pulled downwards by a force of 15 lbs., the extra pound being on account of the friction. To fix our ideas, we shall suppose the 14 lbs. to be raised 1 foot; to lift this load directly, without the intervention of the pulley, 14 foot-pounds would be necessary, but when it is raised by means of the pulley, 15,

foot-pounds are necessary. Hence there is an absolute loss of $\frac{1}{15}$th of the energy when the pulley is used. If a steam-engine of 1 horse-power were employed in raising weights by a rope passing over a pulley similar to that on which we have experimented, only $\frac{14}{15}$ths of the work would be usefully employed; but we find

$$33,000 \times \frac{14}{15} = 30,800.$$

The engine would therefore perform 30,800 foot-pounds of useful work per minute.

180. The effect of friction on a pulley, or on any other machine, is always to waste energy. To perform a piece of work directly requires a certain number of foot-pounds, while to do it by a machine requires more, on account of the loss by friction. This may at first sight appear somewhat paradoxical, as it is well known that, by levers, pulleys, &c., an enormous mechanical advantage may be gained. This subject will be fully explained in the next and following lectures, which relate to the mechanical powers.

181. We shall conclude with a few observations on a point of the greatest importance. We have seen a case where 15 foot-pounds of energy only accomplished 14 foot-pounds of work, and thus 1 foot-pound appeared to be lost. We say that this was expended upon the friction; but what is the friction? The axle is gradually worn away by rubbing in its bearings, and, if it be not properly oiled, it becomes heated. The amount of energy that seems to disappear is partly expended in grinding down the axle, and is partly transformed into heat; it is thus not really lost, but unfortunately assumes a form which we do not require and in which it is rather injurious than otherwise. Indeed we know that energy cannot be destroyed, however it may be

H

transformed; if it disappear in one shape, it is only to reappear in another. A so-called loss of energy by friction only means a diversion of a part of the work to some purpose other than that which we wish to accomplish. It has long been known that matter is indestructible: it is now equally certain that the same may be asserted of energy.

LECTURE VII.

THE PULLEY-BLOCK.

Introduction.—The Single Moveable Pulley.—The Three-sheave Pulley-block.—The Differential Pulley-block.—The Epicycloidal Pulley-block.

INTRODUCTION.

182. In the first lecture I showed how a large weight could be raised by a smaller weight, and I stated that this subject would again occupy our attention. I now fulfil this promise. The questions to be discussed involve the most advantageous methods of employing a small force to overcome a greater. Here is a subject of practical importance. A man of average strength cannot raise more than a hundredweight without great exertion, yet the weights which it is necessary to lift and move about often weigh many hundredweights, or even many tons. It is not always practicable to employ numerous hands for the purpose, nor is a steam-engine or other great source of power at all times available. But what are called the mechanical powers enable the forces at our disposal to be greatly increased. One man, by their aid, can exert as much force as several

could without such assistance; and when they are employed to augment the power of several men or of a steam-engine, gigantic weights, amounting sometimes to hundreds of tons, can be managed with facility.

183. In the various arts we find innumerable cases where great resistances have to be overcome; we also find a corresponding number and variety of devices contrived by human skill to conquer them. The girders of an iron bridge have to be lifted up to their piers; the boilers and engines of an ocean steamer have to be placed in position; a great casting has to be raised from its mould; a railway locomotive has to be placed on the deck of a vessel for transit; a weighty anchor has to be lifted from the bottom of the sea; an iron plate has to be rolled or cut or punched: for all of these cases suitable arrangements must be devised in order that the requisite power may be obtained.

184. We know but little of the means which the ancients employed in raising the vast stones of those buildings which travellers in the East have described to us. It is sometimes thought that a large number of men could have transported these stones without the aid of appliances which we would now use for a similar purpose. But it is more likely that some of the mechanical powers were used, as, with a multitude of men, it is difficult to ensure the proper application of their united strength. In Easter Island, hundreds of miles distant from civilised land, and now inhabited by savages, vast idols of stone have been found in the hills which must have been raised by human labour. It is useless to speculate on the extinct race by whom this work was achieved, or on the means they employed.

185. The mechanical powers are usually enumerated as follows:—The pulley, the lever, the wheel and axle, the wedge, the inclined plane, the screw. These different powers are so

VII.] THE SINGLE MOVEABLE PULLEY. 101

frequently used in combination that the distinctions cannot be always maintained. The classification will, however, suffice to give a general notion of the subject.

186. Many of the most valuable mechanical powers are machines in which ropes or chains play an important part. Pulleys are usually employed wherever it is necessary to change the direction of a rope or chain which is transmitting power. In the present lecture we shall examine the most important mechanical powers that are produced by the combination of pulleys.

THE SINGLE MOVEABLE PULLEY.

187. We commence with the most simple case, that of the single moveable pulley (Fig. 35). The rope is firmly secured at one end A; it then passes down under the moveable pulley B, and upwards over a fixed pulley. To the free end, which depends from the fixed pulley, the power is applied while the load to be raised is suspended from the moveable pulley. We shall first study the relation between the power and the load in a simple way, and then we shall describe a few exact experiments.

188. When the load is raised the moveable pulley itself must of course be also raised, and a part of the power is expended for this purpose. But we can eliminate the weight of the moveable pulley, so far as our calculations are concerned, by first attaching to the power end of the rope a sufficient weight to lift up the moveable pulley when not carrying a load. The weight necessary for doing this is found by trial to be a little over 1·5 lbs. So that when a load is being raised we must reduce the apparent power by 1·5 lbs. to obtain the power really effective.

189. Let us suspend 14 lbs. from the load hook at B, and ascertain what power will raise the load. We leave the weight

of the moveable pulley and 1·5 lbs. of the power at C out of consideration. I then find by experiment that 7 lbs. of effective power is not sufficient to raise the load, but if one pound more be added, the power descends, and the load is raised.

Fig. 35.

Here, then, is a remarkable result; a weight of 8 lbs. has overcome 14 lbs. In this we have the first application of the mechanical powers to increase our available forces.

190. Let us examine the reason of this mechanical advantage. If the load be raised one foot, it is plain that the power must descend two feet: for in order to raise the

VII.]   THE SINGLE MOVEABLE PULLEY.   103

load the two parts of the rope descending to the moveable pulley must each be shortened one foot, and this can only be done by the power descending two feet. Hence when the load of 14 lbs. is lifted by the machine, for every foot it is raised the power must descend two feet : this simple point leads to a conception of the greatest importance, on which depends the efficiency of the pulley. In the study of the mechanical powers it is essential to examine the number of feet through which the power must act in order to raise the load one foot : this number we shall always call the *velocity ratio*.

191. To raise 14 lbs. one foot requires 14 foot-pounds of energy. Hence, were there no such thing as friction, 7 lbs. on the power hook would be sufficient to raise the load; because 7 lbs. descending through two feet yields 14 foot-pounds. But there is a loss of energy on account of friction, and a power of 7 lbs. is not sufficient : 8 lbs. are necessary. Eight lbs. in descending two feet performs 16 foot-pounds ; of these only 14 are utilised on the load, the remainder being the quantity of energy that has been diverted by friction. We learn, then, that in the moveable pulley the quantity of *energy* employed is really greater than that which would lift the weight directly, but that the actual *force* which has to be exerted is less.

192. Suppose that 28 lbs. be placed on the load hook, a few trials assure us that a power of 16 lbs. (but not less) will be sufficient for motion; that is to say, when the load is doubled, we find, as we might have expected, that the power must be doubled also. It is easily seen that the loss of energy by friction then amounts to 4 foot-pounds. We thus verify, in the case of the moveable pulley, the approximate law that the *friction is proportional to the load*.

193. By means of a moveable pulley a man is able to raise a weight nearly double as great as he could lift

directly. From a series of careful experiments it has been found that when a man is employed in the particular exertion necessary for raising weights over a pulley, he is able to work most efficiently when the pull he is required to make is about 40 lbs. A man could, of course, exert greater force than this, but in an ordinary day's work he is able to perform more foot-pounds when the pull is 40 lbs. than when it is larger or smaller. If therefore the weights to be lifted amount to about 80 lbs., energy may really be economized by the use of the single moveable pulley, although by so doing a greater quantity of energy would be actually expended than would have been necessary to raise the weights directly.

194. Some experiments on larger loads have been tried with the moveable pulley we have just described; the results are recorded in Table IX.

TABLE IX.—SINGLE MOVEABLE PULLEY.

Moveable pulley of cast iron $3''\cdot25$ diameter, groove $0''\cdot6$ wide, wrought iron axle $0''\cdot6$ diameter; fixed pulley of cast iron $5''$ diameter, groove $0''\cdot4$ wide, wrought iron axle $0''\cdot6$ diameter, axles oiled; flexible plaited rope $0''\cdot25$ diameter; velocity ratio 2, mechanical efficiency 1·8, useful effect 90 per cent.; formula $P = 2\cdot21 + 0\cdot5453\ R$.

Number of Experiment.	R. Load in lbs.	Observed power in lbs.	P. Calculated power in lbs.	Discrepancies between observed and calculated powers.
1	28	17·5	17·5	0·0
2	57	33·5	33·3	− 0·2
3	85	48·5	48·6	+ 0·1
4	113	64·0	63·8	− 0·2
5	142	80·0	79·6	− 0·4
6	170	94·5	94·9	+ 0·4
7	198	110·5	110·2	− 0·3
8	226	125·5	125·5	0·0

The dimensions of the pulleys are precisely stated because, for pulleys of different construction, the numerical

VII.] THE SINGLE MOVEABLE PULLEY. 105

coefficients would not necessarily be the same. An attentive study of this table will, however, show the general character of the relation between the power and the load in all arrangements of this class.

The table consists of five columns. The first contains merely the numbers of the experiments for convenience of reference. In the second column, headed R, the loads, expressed in pounds, which are raised in each experiment, are given; that is, the weight attached to the hook, not including the weight of the lower pulley. The weight of this pulley is not included in the stated loads. In the third column the powers are recorded, which were found to be sufficient to raise the corresponding loads in the second column. Thus, in experiment 7, it is found that a power of 110·5 lbs. will be sufficient to raise a load of 198 lbs. The third column has thus been determined by gradually increasing the power until motion begins.

195. From an examination of the columns showing the power and the load, we see that the power always amounts to more than half the load. The excess is partly due to a small portion of the power (about 1·5 lbs.) being employed in raising the lower block, and partly to friction. For example, in experiment 7, if there had been no friction and if the lower block were without weight, a power of 99 lbs. would have been sufficient; but, owing to the presence of these disturbing causes, 110·5 lbs. are necessary: of this amount 1·5 lbs. is due to the weight of the pulley, 10 lbs. is the force of friction, and the remaining 99 lbs. raises the load.

196. By a calculation based on this table we have ascertained a certain relation between the power and the load; they are connected by the formula which may be enunciated as follows :

The power is found by multiplying the weight of the load

into 0·5453, and adding 2·21 to the product. Calling P the power and R the load, we may express the relation thus: $P = 2·21 + 0·5453\ R$. For example, in experiment 5, the product of 142 and 0·5453 is 77·43, to which, when 2·21 is added, we find for P 79·64, very nearly the same as 80 lbs., the observed value of the power.

In the fourth column the values of P calculated by means of this formula are given, and in the last we exhibit the discrepancies between the observed and the calculated values for the sake of comparison. It will be seen that the discrepancy in no case amounts to 0·5 lb., consequently the formula expresses the experiments very well. The mode of deducing it is given in the Appendix.

197. The quantity 2·21 is partly that portion of the power expended in overcoming the weight of the moveable pulley, and partly arises from friction.

198. We can readily calculate from the formula how much power will be required to raise a given weight; for example, suppose 200 lbs. be attached to the moveable pulley, we find that 111 lbs. must be applied as the power. But in order to raise 200 lbs. one foot, the power exerted must act over two feet; hence the number of foot-pounds required is 2 × 111 = 222. The quantity of energy that is lost is 22 foot-pounds. Out of every 222 foot-pounds applied, 200 are usefully employed; that is to say, about 90 per cent. of the applied energy is utilized, while the remaining 10 per cent. is lost by friction.

THE THREE-SHEAVE PULLEY-BLOCK.

199. The next arrangement we shall employ is a pair of pulley-blocks S T, Fig. 35, each containing three sheaves, as the small pulleys are termed. A rope is fastened to the upper block, S; it then passes down to the lower block T

VII.] THE THREE-SHEAVE PULLEY-BLOCK. 107

under one sheave, up again to the upper block and over a sheave, and so on, as shown in the figure. To the end of the rope from the last of the upper sheaves the power H is applied, and the load G is suspended from the hook attached to the lower block. When the rope is pulled, it gradually raises the lower block; and to raise the load one foot, each of the six parts of the rope from the upper block to the lower block must be shortened one foot, and therefore the power must have pulled out six feet of rope. Hence, for every foot that the load is raised the power must have acted through six feet; that is to say, the *velocity ratio* is 6.

200. If there were no friction, the power would only be one-sixth of the load. This follows at once from the principles already explained. Suppose the load be 60 lbs., then to raise it one foot would require 60 foot-pounds; and the power must therefore exert 60 foot-pounds; but the power moves over six feet, therefore a power of 10 lbs. would be sufficient. Owing, however, to friction, some energy is lost, and we must have recourse to experiment in order to test the real efficiency of the machine. The single moveable pulley nearly doubled our power; we shall prove that the three-sheave pulley-block will quadruple it. In this case we deal with larger weights, with reference to which we may leave the weight of the lower block out of consideration.

201. Let us first attach 1 cwt. to the load hook; we find that 29 lbs. on the power hook is the smallest weight that can produce motion: this is only 1 lb. more than one-quarter of the load raised. If 2 cwt. be the load, we find that 56 lbs. will just raise it: this time the power is exactly one-quarter of the load. The experiment has been tried of placing 4 cwt. on the hook; it is then found that 109 lbs. will raise it, which is only 3 lbs. short of 1 cwt. These experiments demonstrate that for a three-sheave pulley-

block of this construction we may safely apply th rule, that *the power is one-quarter of the load.*

202. We are thus enabled to see how much of our exertion in raising weights must be expended in merely overcoming friction, and how much may be utilized. Suppose for example that we have to raise a weight of 100 lbs. one foot by means of the pulley-block; the power we must apply is 25 lbs., and six feet of rope must be drawn out from between the pulleys: therefore the power exerts 150 foot-pounds of energy. Of these only 100 foot-pounds are usefully employed, and thus 50 foot-pounds, one-third of the whole, have been expended on friction. Here we see that notwithstanding a small force overcomes a large one, there is an actual loss of energy in the machine. The real advantage of course is that by the pulley-block I can raise a greater weight than I could move without assistance, but I do not create energy; I merely modify it, and lose by the process.

203. The result of another series of experiments made with this pair of pulley-blocks is given in Table X.

TABLE X.—THREE-SHEAVE PULLEY-BLOCKS.

Sheaves cast iron $2''\cdot5$ diameter; plaited rope $0'''\cdot25$ diameter; velocity ratio 6; mechanical advantage 4; useful effect 67 per cent.; formula $P = 2\cdot36 + 0\cdot238\ R$.

Number of Experiment.	R. Load in lbs.	Observed power in lbs.	P. Calculated power in lbs.	Discrepancies between observed and calculated power.
1	57	15.5	15.9	+ 0.4
2	114	29.5	29.5	0.0
3	171	43.5	43.1	− 0.4
4	228	56.0	56.6	+ 0.6
5	281	70.0	69.2	− 0.8
6	338	83.0	82.8	− 0.2
7	395	97.0	96.4	− 0.6
8	452	109.0	109.9	+ 0.9

VII.] THE THREE SHEAVE PULLEY-BLOCK. 109

204. This table contains five columns; the weights raised (shown in the second column) range up to somewhat over 4 cwt. The observed values of the power are given in the third column; each of these is generally about one-quarter of the corresponding value of the load. There is, however, a more accurate rule for finding the power; it is as follows.

205. To find the power necessary to raise a given load, multiply the loads in lbs. by 0·238, and add 2·36 lbs. to the product. We may express the rule by the formula $P = 2·36 + 0·238\ R$.

206. To find the power which would raise 228 lbs.; the product of 228 and 0·238 is 54·26; adding 2·36, we find 56·6 lbs. for the power required; the actual observed power is 56 lbs., so that the rule is accurate to within about half a pound. In the fourth column will be found the values of P calculated by means of this rule. In the fifth column, the discrepancies between the observed and the calculated values of the powers are given, and it will be seen that the difference in no case reaches 1 lb. Of course it will be understood that this formula is only reliable for loads which lie between those employed in the first and last of the experiments. We can calculate the power for any load between 57 lbs. and 452 lbs., but for loads much larger than 452 or less than 57 it would probably be better to use the simple fourth of the load rather than the power computed by the formula.

207. I will next perform an experiment with the three-sheave pulley-block, which will give an insight into the exact amount of friction without calculation by the help of the velocity ratio. We first counterpoise the weight of the lower block by attaching weights to the power. It is found that about 1·6 lbs. is sufficient for this purpose. I attach a 56 lb. weight as a load, and find that 13·1 lbs. is sufficient power for motion. This amount is partly com-

posed of the force necessary to raise the load if there were no friction, and the rest is due to the friction. I next gradually remove the power weights: when I have taken off a pound, you see the power and the load balance each other; but when I have reduced the power so low as 5·5 lbs. (not including the counterpoise for the lower block), the load is just able to overhaul the power, and run down. We have therefore proved that a power of 13·1 lbs. or greater raises 56 lbs., that any power between 13·1 lbs. and 5·5 lbs. balances 56 lbs., and that any power less than 5·5 lbs. is raised by 56 lbs.

When the power is raised, the force of friction, together with the power, must be overcome by the load. Let us call X the real power that would be necessary to balance 56 lbs. in a perfectly frictionless machine, and Y the force of friction. We shall be able to determine X and Y by the experiments just performed. When the load is raised a power equal to $X + Y$ must be applied, and therefore $X + Y = 13\cdot1$. On the other hand, when the power is raised, the force X is just sufficient to overcome both the friction Y and the weight 5·5; therefore $X = Y + 5\cdot5$.

Solving this pair of equations, we find that $X = 9\cdot3$ and $Y = 3\cdot8$. Hence we infer that the power in the frictionless machine would be 9·3; but this is exactly what would have been deduced from the velocity ratio, for $56 \div 6 = 9\cdot3$ lbs. In this result we find a perfect accordance between theory and experiment.

THE DIFFERENTIAL PULLEY-BLOCK.

208. By increasing the number of sheaves in a pair of pulley-blocks the power may be increased; but the length of rope (or chain) requisite for several sheaves becomes a practical inconvenience. There are also other reasons

VII.] THE DIFFERENTIAL PULLEY-BLOCK. 111

which make the differential pulley-block, which we shall now consider, more convenient for many purposes than the common pulley blocks when a considerable augmentation of power is required.

209. The principle of the differential pulley is very ancient, and in modern times it has been embodied in a machine of practical utility. The object is to secure, that while the power moves over a considerable distance, the load shall only be raised a short distance. When this has been attained, we then know by the principle of energy that we have gained a mechanical advantage.

210. Let us consider the means by which this is effected in that ingenious contrivance, Weston's differential pulley-block. The principle of this machine will be understood from Fig. 36 and Fig. 37.

It consists of three parts,—an upper pulley-block, a moveable pulley, and an endless chain. We shall briefly describe them. The upper block P is furnished with a hook for attachment to a support. The sheave it contains resembles two sheaves, one a little smaller than the other, fastened together: they are in fact one piece. The grooves are provided with ridges, adapted to prevent the chain from slipping. The lower pulley Q consists of one sheave, which is also furnished with a groove; it carries a hook, to

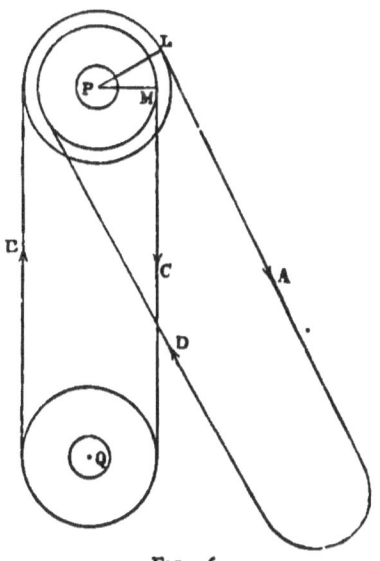

FIG. 36.

which the load is attached. The endless chain performs a part that will be understood from the sketch of the principle in Fig. 36. The chain passes from the hand at A up to L over the larger groove in the upper pulley, then downwards at B, under the lower pulley, up again at C, over the smaller groove in the upper pulley at A, and then back again by D to the hand at A. When the hand pulls the chain downwards, the two grooves of the upper pulley begin to turn together in the direction shown by the arrows on the chain. The large groove is therefore winding up the chain, while the smaller groove is lowering.

211. In the pulley which has been employed in the experiments to be described, the effective circumference of the large groove is found to be $11''\cdot 84$, while that of the small groove is $10''\cdot 36$. When the upper pulley has made one revolution, the large groove must have drawn up $11''\cdot 84$ of chain, since the chain cannot slip on account of the ridges; but in the same time the small groove has lowered $10''\cdot 36$ of chain: hence when the upper pulley has revolved once, the chain between the two must have been shortened by the difference between $11''\cdot 84$ and $10''\cdot 36$, that is by $1''\cdot 48$; but this can only have taken place by raising the moveable pulley through half $1''\cdot 48$, that is, through a space $0''\cdot 74$. The power has then acted through $11''\cdot 84$, and has raised the resistance $0''\cdot 74$. The power has therefore moved through a space 16 times greater than that through which the load moves. In fact, it is easy to verify by actual trial that the power must be moved through 16 feet in order that the load may be raised 1 foot. We express this by saying that the *velocity ratio* is 16.

212. By applying power to the chain at D proceeding from the smaller groove, the chain is lowered by the large groove faster than it is raised by the small one, and the lower

VII.] THE DIFFERENTIAL PULLEY-BLOCK. 113

pulley descends. The load is thus raised or lowered by simply pulling one chain A or the other D.

213. We shall next consider the mechanical efficiency of the differential pulley-block. The block (Fig. 37) which we shall use is intended to be worked by one man, and will raise any weight not exceeding a quarter of a ton.

We have already learned that with this block the power must act through sixteen feet for the load to be raised one foot. Hence, were it not for friction, the power need only be the sixteenth part of the load. A few trials will show us that the real efficiency is not so large, and that in fact more than half the work exerted is merely expended upon overcoming friction. This will lead afterwards to a result of considerable practical importance.

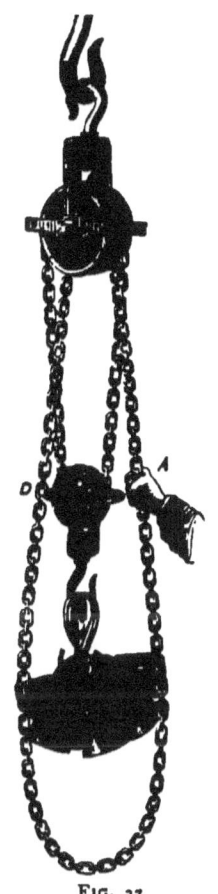

FIG. 37.

214. Placing upon the load-hook a weight of 200 lbs., I find that 38 lbs. attached to a hook fastened on the power-chain is sufficient to raise the load; that is to say, the power is about one-sixth of the load. If I make the load 400 lbs. I find the requisite power to be 64 lbs., which is only about 3 lbs. less than one-sixth of 400 lbs. We may safely adopt the practical rule, that with this differential pulley-block a man would be able to raise a weight six times as great as he could raise without such assistance.

215. A series of experiments carefully tried with different loads have given the results shown in Table XI.

TABLE XI.—THE DIFFERENTIAL PULLEY-BLOCK.

Circumference of large groove 11"·84, of small groove 10"·36; velocity ratio 16; mechanical efficiency 6·07; useful effect 38 per cent; formula $P = 3·87 + 0·1508 R$.

Number of Experiment.	R. Load in lbs.	Observed power in lbs.	P. Calculated power in lbs.	Difference of the observed and calculated values.
1	56	10	12·3	+ 2·3
2	112	20	20·8	+ 0·8
3	168	31	29·2	− 1·8
4	224	38	37·7	− 0·3
5	280	48	46·1	− 1·9
6	336	54	54·6	+ 0·6
7	392	64	63·1	− 0·9
8	448	72	71·5	− 0·5
9	504	80	80·0	0·0
10	560	86	88·4	+ 2·4

The first column contains the numbers of the experiments, the second the weights raised, the third the observed values of the corresponding powers. From these the following rule for finding the power has been obtained:—

216. To find the power, multiply the load by 0·1508, and add 3·87 lbs. to the product; this rule may be expressed by the formula $P = 3·87 + 0·1508 R$. (See Appendix.)

217. The calculated values of the powers are given in the fourth column, and the differences between the observed and calculated values in the last column. The differences do not in any case amount to 2·5 lbs., and considering that the loads raised are up to a quarter of a ton, the formula represents the experiments with satisfactory precision.

218. Suppose for example 280 lbs. is to be raised; the product of 280 and 0·1508 is 42·22, to which, when 3·87 is

VII.]   THE DIFFERENTIAL PULLEY-BLOCK.   115

added, we find 46·09 to be the requisite power. The mechanical efficiency found by dividing 46·09 into 280 is 6·07.

219. To raise 280 lbs. one foot 280 foot-pounds of energy would be necessary, but in the differential pulley-block 46·09 lbs. must be exerted for a distance of 16 feet in order to accomplish this object. The product of 46·09 and 16 is 737·4. Hence the differential pulley-block requires 737·4 foot-pounds of energy to be applied in order to yield 280 useful foot-pounds; but 280 is only 38 per cent. of 737·4, and therefore with a load of 280 lbs. only 38 per cent. of the energy applied to a differential pulley-block is utilized. In general, we may state that not more than about 40 per cent. is profitably used, and that the remainder is expended in overcoming friction.

220. It is a remarkable and useful property of the differential pulley, that a weight which has been hoisted will remain suspended when the hand is removed, even though the chain be not secured in any manner. The pulleys we have previously considered do not possess this convenient property. The weight raised by the three-sheave pulley-block, for example, will run down unless the free end of the rope be properly secured. The difference in this respect between these two mechanical powers is not a consequence of any special mechanism; it is simply caused by the excessive friction in the differential pulley-block.

221. The reason why the load does not run down in the differential pulley may be thus explained. Let us suppose that a weight of 400 lbs. is to be raised one foot by the differential pulley-block; 400 units of work are necessary, and therefore 1,000 units of work must be applied to the

power chain to produce the 400 units (since only 40 per cent. is utilized). The friction will thus have consumed 600 units of work when the load has been raised one foot. If the power-weight be removed, the pressure supported by the upper pulley-block is diminished. In fact, since the power-weight is about $\frac{1}{5}$th of the load, the pressure on the axle when the power-weight has been removed is only $\frac{6}{7}$ths of its previous value. The friction is nearly proportional to that pressure: hence when the power has been removed the friction on the upper axle is $\frac{6}{7}$ths of its previous value, while the friction on the lower pulley remains unaltered.

We may therefore assume that the total friction is at least $\frac{6}{7}$ths of what it was before the power-weight was removed. Will friction allow the load to descend? 600 foot-pounds of work were required to overcome the friction in the ascent: at least $\frac{6}{7} \times 600 = 514$ foot-pounds would be necessary to overcome friction in the descent. But where is this energy to come from? The load in its descent could only yield 400 units, and thus descent by the mere weight of the load is impossible. To enable the load to descend we have actually to aid the movement by pulling the chain D (Figs. 36 and 37), which proceeds from the small groove in the upper pulley.

222. The principle which we have here established extends to other mechanical powers, and may be stated generally. *Whenever more than half the applied energy is consumed by friction, the load will remain without running down when the machine is left free.*

THE EPICYCLOIDAL PULLEY-BLOCK.

223. We shall conclude this lecture with some experiments upon a useful mechanical power introduced by Mr. Eade under the name of the epicycloidal pulley-block. It

VII.] THE EPICYCLOIDAL PULLEY-BLOCK. 117

is shown in Fig. 33, and also in Fig. 49. In this machine there are two chains: one a slight endless chain to which the power is applied; the other a stout chain which has a hook at each end, from either of which the load may be suspended. Each of these chains passes over a sheave in the block: these sheaves are connected by an ingenious piece of mechanism which we need not here describe. This mechanism is so contrived that, when the power causes the sheave to revolve over which the slight chain passes, the sheave which carries the large chain is also made to revolve, but very slowly.

224. By actual trial it is ascertained that the power must be exerted through twelve feet and a half in order to raise the load one foot; the velocity ratio of the machine is therefore 12·5.

225. If the machine were frictionless, its mechanical efficiency would be of course equal to its velocity ratio; owing to the presence of friction the mechanical efficiency is less than the velocity ratio, and it will be necessary to make experiments to determine the exact value. I attach to the load-hook a weight of 280 lbs., and insert a few small hooks into the links of the power chain in order to receive weights: 56 lbs. is sufficient to produce motion; hence the mechanical efficiency is 5. Had there been no friction a power of 56 lbs. would have been capable of overcoming a load of $12·5 \times 56 = 700$ lbs. Thus 700 units of energy must be applied to the machine in order to perform 280 units of work. In other words, only 40 per cent. of the applied energy is utilized.

226. An extended series of experiments upon the epicycloidal pulley-block is recorded in Table XII.

TABLE XII.—THE EPICYCLOIDAL PULLEY-BLOCK.

Size adapted for lifting weights up to 5 cwt.; velocity ratio 12·5; mechanical efficiency 5; useful effect 40 per cent.; calculated formula $P = 5\cdot8 + 0\cdot185\ R$.

Number of Experiment.	R. Loads in lbs.	Observed power in lbs.	P. Calculated power in lbs.	Difference of the observed and calculated values.
1	56	15	16·2	+ 1·2
2	112	27	26·5	− 0·5
3	168	40	36·9	− 3·1
4	224	47	47·2	+ 0·2
5	280	56	57·6	+ 1·6
6	336	66	68·0	+ 2·0
7	392	78	78·3	+ 0·3
8	448	88	88·6	+ 0·6
9	504	100	99·0	− 1·0
10	560	110	109·4	− 0·6

The fourth column shows the calculated values of the powers derived from the formula. It will be seen by the last column that the formula represents the experiments with but little error.

227. Since 60 per cent. of energy is consumed by friction, this machine, like the differential pulley-block, sustains its load when the chains are free. The differential pulley-block gives a mechanical efficiency of 6, while the epicycloidal pulley-block has only a mechanical efficiency of 5, and so far the former machine has the advantage; on the other hand, that the epicycloidal pulley contains but one block, and that its lifting chain has two hooks, are practical conveniences strongly in its favour.

LECTURE VIII.

THE LEVER.

The Lever of the First Order.—The Lever of the Second Order.—
The Shears.—The Lever of the Third Order.

THE LEVER OF THE FIRST ORDER.

228. There are many cases in which a machine for overcoming great resistance is necessary where pulleys would be quite inapplicable. To meet these various demands a correspondingly various number of contrivances has been devised. Amongst these the lever in several different forms holds an important place.

229. The lever of the first order will be understood by reference to Fig. 38. It consists of a straight rod, to one end of which the power is applied by means of the weight C. At another point B the load is raised, while at A the rod is supported by what is called the fulcrum. In the case represented in the figure the rod is of iron, $1'' \times 1''$ in section and $6'$ long; it weighs 19 lbs. The power is produced by a 56 lb. weight: the fulcrum consists of a moderately sharp steel edge firmly secured to the framework.

120   EXPERIMENTAL MECHANICS.   [LECT.

The load in this case is replaced by a spring balance H, and the hook of the balance is attached to the frame. The spring is strained by the action of the lever, and the index

FIG. 38.

records the magnitude of the force produced at the short end. This is the lever with which we shall commence our experiments.

VIII.] THE LEVER OF THE FIRST ORDER. 121

230. In examining the relation between the power and the load, the question is a little complicated by the weight of the lever itself (19 lbs.), but we shall be able to evade the difficulty by means similar to those employed on a former occasion (Art. 60); we can counterpoise the weight of the iron bar. This is easily done by applying a hook to the middle of the bar at D, thence carrying a rope over a pulley F, and suspending a weight G of 19 lbs. from its free extremity. Thus the bar is balanced, and we may leave its weight out of consideration.

231. We might also adopt another plan analogous to that of Art. 51, which is however not so convenient. The weight of the bar produces a certain strain upon the spring balance. I may first read off the strain produced by the bar alone, and then apply the weight C and read again. The observed strain is due both to the weight C and to the weight of the bar. If I subtract the known effect of the bar, the remainder is the effect of C. It is, however, less complicated to counterpoise the bar, and then the strains indicated by the balance are entirely due to the power.

232. The lever is 6' long; the point E is 6" from the end, and BC is 5' long. BC is divided into 5 equal portions of 1'; A is at one of these divisions, 1' distant from B, and C is 5' distant from B in the figure; but C is capable of being placed at any position, by simply sliding its ring along the bar.

233. The mode of experimenting is as follows:—The weight is placed on the bar at the position C: a strain is immediately produced upon H; the spring stretches a little, and the bar becomes inclined. It may be noticed that the hook of the spring balance passes through the eye of a wirestrainer, so that by a few turns of the nut upon the strainer the lever can be restored to the horizontal position.

234. The power of 56 lbs. being 4' from the fulcrum,

while the load is 1' from the fulcrum, it is found that the strain indicated by the balance is 224 lbs.; that is, four times the amount of the power. If the weight be moved, so as to be 3' from the fulcrum, the strain is observed to be 168 lbs.; and whatever be the distance of the power from the fulcrum, we find that the strain produced is obtained by multiplying the magnitude of the power in pounds by the distance expressed in feet, and fractional parts of a foot. This law may be expressed more generally by stating that *the power is to the load as the distance of the load from the fulcrum is to the distance of the power from the fulcrum.*

235. We can verify this law under varied circumstances. I move the steel edge which forms the fulcrum of the lever until the edge is 2' from B, and secure it in that position. I place the weight C at a distance of 3' from the fulcrum. I now find that the strain on the balance is 84 lbs.; but 84 is to 56 as 3 is to 2, and therefore the law is also verified in this instance.

236. There is another aspect in which we may express the relation between the power and the load. The law in this form is thus stated: *The power multiplied by its distance from the fulcrum is equal to the load multiplied by its distance from the fulcrum.* Thus, in the case we have just considered, the product of 56 and 3 is 168, and this is equal to the product of 84 and 2. The distances from the fulcrum are commonly called the arms of the lever, and the rule is expressed by stating that *The power multiplied into its arm is equal to the load multiplied into its arm:* hence the load may be found by dividing the product of the power and the power arm by the load arm. This simple law gives a very convenient method of calculating the load, when we know the power and the distances of the power and the load from the fulcrum.

237. When the power arm is longer than the load arm, the load is greater than the power; but when the power arm is shorter than the load arm, the power is greater than the load.

We may regard the strain on the balance as a power which supports the weight, just as we regard the weight to be a power producing the strain on the balance. We see, then, that for the lever of the first order to be efficient as a mechanical power it is necessary that the power arm be longer than the load arm.

238. The lever is an extremely simple mechanical power; it has only one moving part. Friction produces but little effect upon it, so that the laws which we have given may be actually applied in practice, without making any allowance for friction. In this we notice a marked difference between the lever and the pulley-blocks already described.

239. In the lever of the first order we find an excellent machine for augmenting power. A power of 14 lbs. can by its means overcome a resistance of a hundredweight, if the power be eight times as far from the fulcrum as the load is from the fulcrum. This principle it is which gives utility to the crowbar. The end of the bar is placed under a heavy stone, which it is required to raise; a support near that end serves as a fulcrum, and then a comparatively small force exerted at the power end will suffice to elevate the stone.

240. The applications of the lever are innumerable. It is used not only for increasing power, but for modifying and transforming it in various ways. The lever is also used in weighing-machines, the principles of which will be readily understood, for they are consequences of the law we have explained. Into these various appliances it is not our intention to enter at present; the great majority of them

may, when met with, be easily understood by the principle we have laid down.

THE LEVER OF THE SECOND ORDER.

241. In the lever of the second order the power is at one end, the fulcrum at the other end, and the load lies between the two : this lever therefore differs from the lever of the first order, in which the fulcrum lies between the two forces. The relation between the power and the load in the lever of the second order may be studied by the arrangement in Fig. 39.

242. The bar A C is the same rod of iron $72'' \times 1'' \times 1''$ which was used in the former experiment. The fulcrum A is a steel edge on which the bar rests ; the power consists of a spring balance H, in the hook of which the end C of the bar rests ; the spring balance is sustained by a wire-strainer, by turning the nut of which the bar may be adjusted horizontally. The part of the bar between the fulcrum A and the power C is divided into five portions, each 1′ long, and the points A and C are each 6″ distant from the extremities of the bar. The load employed is 56 lbs. ; through the ring of this weight the bar passes, and thus the bar supports the load. The bar is counterpoised by the weight of 19 lbs. at G, in the manner already explained (Art. 231).

243. The mode of experimenting is as follows :—Let the weight B be placed 1′ from the fulcrum ; the strain shown by the spring balance is about 11 lbs. If we calculate the value of the power by the rule already given, we should have found the same result. The product of the load by its distance from the fulcrum is 56, the distance of the power from the fulcrum is 5 ; hence the value of the power should be $56 \div 5 = 11\cdot2$.

244. If the weight be placed 2′ from the fulcrum the

VIII.] THE LEVER OF THE SECOND ORDER. 125

strain is about 22·5 lbs. and it is easy to ascertain that this is the same amount as would have been found by the application of the rule. A similar result would have been

Fig. 39.

obtained if the 56 lb. weight had been placed upon any other part of the bar; and hence we may regard the rule proved for the lever of the second order as well as for the lever of the first order: that, the power multiplied by its distance from the fulcrum is equal to the load multiplied by its distance from the fulcrum. In the present case the load is uniformly 56 lbs., while the power by which it is sustained is always less than 56 lbs.

FIG. 40.

245. The lever of the second order is frequently applied to practical purposes; one of the most instructive of these applications is illustrated in the shears shown in Fig. 40.

The shears consist of two levers of the second order, which by their united action enable a man to exert a greatly increased force, sufficient, for example, to cut with ease a rod of iron 0″·25 square. The mode of action is simple. The first lever A F has a handle at one end F, which is 22″ distant from the other end A, where the fulcrum is placed.

VIII.] THE LEVER OF THE SECOND ORDER. 127

At a point B on this lever, $1''\cdot 8$ distant from the fulcrum A, a short link B C is attached; the end of the link C is jointed to a second lever C D; this second lever is $8''$ long; it forms one edge of the cutting shears, the other edge being fixed to the framework.

246. I place a rod of iron $0''\cdot 25$ square between the jaws of the shears in the position E, the distance D E being $3''\cdot 5$, and proceed to cut the iron by applying pressure to the handle. Let us calculate the amount by which the levers increase the power exerted upon F. Suppose for example that I press downwards on the handle with a force of 10 lbs., what is the magnitude of the pressure upon the piece of iron? The effect of each lever is to be calculated separately. We may ascertain the power exerted at B by the rule of moments already explained; the product of the power and its arm is $22 \times 10 = 220$: this divided by the number of inches, $1\cdot 8$ in the line A B, gives a quotient 122, and this quotient is the number of pounds pressure which is exerted by means of the link upon the second lever. We proceed in the same manner to find the magnitude of the pressure upon the iron at E. The product of 122 and 8 is 976. This is divided by $3\cdot 5$, and the quotient found is 279. Hence the exertion of a pressure of 10 lbs. at F produces a pressure of 279 lbs. at E. In round numbers, we may say that the pressure is magnified 28-fold by means of this combination of levers of the second order.

247. A pressure of 10 lbs. is not sufficient to shear across the bar of iron, even though it be magnified to 279 lbs. I therefore suspend weights from F, and gradually increase the load until the bar is cut. I find at the first trial that 112 lbs. is sufficient, and a second trial with the same bar gives 114 lbs.; 113 lbs., the mean between these results, may be considered an adequate force. This is the load on

F; the real pressure on the bar is $113 \times 27\cdot9 = 3153$ lbs.: thus the actual pressure which was necessary to cut the bar amounted to more than a ton.

248. We can calculate from this experiment the amount of force necessary to shear across a bar one square inch in section. We may reasonably suppose that the necessary power is proportional to the section, and therefore the power will bear to 3153 lbs. the proportion which a square of one inch bears to the square of a quarter inch; but this ratio is 16: hence the force is 16×3153 lbs., equal to about 22·5 tons.

249. It is noticeable that 22·5 tons is nearly the force which would suffice to tear the bar in sunder by actual tension. We shall subsequently return to the subject of shearing iron in the lecture upon Inertia (Lecture XVI.).

THE LEVER OF THE THIRD ORDER.

250. The lever of the third order may be easily understood from Fig. 39, of which we have already made use. In the lever of the third order the fulcrum is at one end, the load is at the other end, while the power lies between the two. In this case, then, the power is represented by the 56 lb. weight, while the load is indicated by the spring balance. The power always exceeds the load, and consequently this lever is to be used where speed is to be gained instead of power. Thus, for example; when the power, 56 lbs., is 2' distant from the fulcrum, the load indicated by the spring balance is about 23 lbs.

251. The treadle of a grindstone is often a lever of the third order. The fulcrum is at one end, the load is at the other end, and the foot has only to move through a small distance.

VIII.] THE LEVER OF THE THIRD ORDER. 129

252. The principles which have been discussed in Lecture III. with respect to parallel forces explain the laws now laid down for levers of different orders, and will also enable us to express these laws more concisely.

253. A comparison between Figs. 20 and 39 shows that the only difference between the contrivances is that in Fig. 20 we have a spring balance C in the same place as the steel edge A in Fig 39. We may in Fig. 20 regard one spring balance as the power, the other as the fulcrum, and the weight as the load. Nor is there any essential difference between the apparatus of Fig. 38 and that of Fig. 20. In Fig. 38 the bar is pulled down by a force at each end, one a weight, the other a spring balance, while it is supported by the upward pressure of the steel edge. In Fig. 20 the bar is being pulled upwards by a force at each end, and downwards by the weight. The two cases are substantially the same. In each of them we find a bar acted upon by a pair of parallel forces applied at its extremities, and retained in equilibrium by a third force.

254. We may therefore apply to the lever the principles of parallel forces already explained. We showed that two parallel forces acting upon a bar could be compounded into a resultant, applied at a certain point of the bar. We have defined the moment of a force (Art. 64), and proved that the moments of two parallel forces about the point of application of their resultant are equal.

255. In the lever of the first order there are two parallel forces, one at each end; these are compounded into a resultant, and it is necessary that this resultant be applied exactly over the steel edge or fulcrum in order that the bar may be maintained at rest. In the levers of the second and third orders, the power and the load are two parallel forces acting in opposite directions; their resultant, therefore, does

K

not lie between the forces, but is applied on the side of the greater, and at the point where the steel edge supports the bar. In all cases the moment of one of the forces about the fulcrum must be equal to that of the other. From the equality of moments it follows that the product of the power and the distance of the power from the fulcrum equals the product of the load, and the distance of the load from the fulcrum : this principle suffices to demonstrate the rules already given.

256. The laws governing the lever may be deduced from the principle of work; the load, if nearer than the power to the fulcrum, is moved through a smaller distance than the power. Thus, for example, in the lever of the first order: if the load be 12 times as far as the power from the fulcrum, then for every inch the load moves it can be demonstrated that the power must move 12 inches. The number of units of work applied at one end of a machine is equal to the number yielded at the other, always excepting the loss due to friction, which is, however, so small in the lever that we may neglect it. If then a power of 1 lb. be applied to move the power end through 12 inches, one unit of work will have been put into the machine. Hence one unit of work must be done on the load, but the load only moves through $\frac{1}{12}$ of a foot, and therefore a load of 12 lbs. could be overcome : this is the same result as would be given by the rule (Art. 236).

257. To conclude : we have first determined by actual experiment the relation between the power and the load in the lever; we have seen that the law thus obtained harmonizes with the principle of the composition of parallel forces; and, finally, we have shown how the same result can be deduced from the fertile and important principle of work.

LECTURE IX.

THE INCLINED PLANE AND THE SCREW.

The Inclined Plane without Friction.—The Inclined Plane with Friction.—The Screw.—The Screw-jack.—The Bolt and Nut.

THE INCLINED PLANE WITHOUT FRICTION.

258. THE mechanical powers now to be considered are often used for other purposes beside those of raising great weights. For example: the parts of a structure have to be forcibly drawn together, a powerful compression has to be exerted, a mass of timber or other material has to be riven asunder by splitting. For purposes of this kind the inclined plane in its various forms, and the screw, are of the greatest use. The screw also, in the form of the screw-jack, is sometimes used in raising weights. It is principally convenient when the weight is enormously great, and the distance through which it has to be raised comparatively small.

259. We shall commence with the study of the inclined plane. The apparatus used is shown in Fig. 41. A B is a plate of glass 4' long, mounted on a frame and turning

round a hinge at A; B D is a circular arc, with its centre at A, by which the glass may be supported; D C is a vertical rod, to which the pulley C is clamped. This pulley can be moved up and down, to be accommodated to the position of A B; the pulley is made of brass, and turns very freely. A little truck R is adapted to run on the plane of glass.

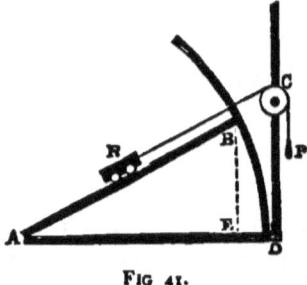

FIG 41.

The truck is laden to weigh 1 lb., and this weight is unaltered throughout the experiments; the wheels are very free, so that the truck runs with but little friction.

260. But the friction, though small, is appreciable, and it will be necessary to measure the amount and then endeavour to counteract its effect upon the motion. The silk cord attached to the truck is very fine, and its weight is neglected. A series of weights is provided; they are made from pieces of brass wire, and weigh 0·1 lb. and 0·01 lb.: these can easily be hooked into the loop on the cord at P. We first make the plane A B horizontal, and bring down the pulley C so that the cord shall be parallel to the plane; we find that a force must be applied by the cord in order to draw the truck along the plane: this force is of course the friction, and by a suitable weight at P the friction may be said to be counterbalanced. But we cannot expect that the friction will be the same when the plane is horizontal as when the plane is inclined. We must therefore examine this question by a method analogous to that used in Art. 207.

261. Let the plane be elevated until B E, the elevation of B above A D, is 20"; let C be properly adjusted: it is found that when P is 0·45 lb. R is just pulled up; and on the

IX.] INCLINED PLANE WITHOUT FRICTION. 133

other hand, when P is only 0·40 lb. the truck descends and raises P; and when P has any value intermediate between these two, the truck remains in equilibrium. Let us denote the force of gravity acting down the plane by R, and it follows that R must be 0·425 lb., and the friction 0·025 lb. For when P raises R, it must overcome friction as well as R; therefore the power must be 0·025 + 0·425 = 0·45. On the other hand, when R raises P, it must also overcome the friction 0·025, therefore P can only be 0·425 − 0·025 = 0·40; and R is thus found to be a mean between the greatest and least values of P consistent with equilibrium. If the plane be raised so that the height B E is 33", the greatest and least values of P are 0·66 and 0·71; therefore R is 0·685 and the friction 0·025, the same as before. Finally, making the height B E only 2", the friction is found to be 0·020, which is not much less than the previous determinations. These experiments show that we may consider this very small friction to be practically constant at these inclinations. (Were the friction large, other methods are necessary, see Art. 265.) As in the experiments R is always *raised* we shall give P the permanent load of 0·025 lb., thus sufficiently counteracting friction, which we may therefore dismiss from consideration. It is hardly necessary to remark that, in afterwards recording the weights placed at P, this counterpoise is not to be included.

262. We have now the means of studying the relation between the power and the load in the frictionless inclined plane. The incline being set at different elevations, we shall observe the force necessary to draw up the constant load of 1 lb. Our course will be guided by first making use of the principle of energy. Suppose B E to be 2'; when the truck has been moved from the bottom of the plane to the top, it will have been raised vertically through a height of 2',

and two units of energy must have been consumed. But the plane being 4′ long, the force which draws up the truck need only be 0·5 lb., for 0·5 lb. acting over 4′ produces two units of work. In general, if l be the length of the plane and h its height, R the load, and P the power, the number of units of energy necessary to raise the load is Rh, and the number of units expended in pulling it up the plane is Pl: hence $Rh = Pl$, and consequently $P:h::R:l$; that is, the power is to the height of the plane as the load is to its length. In the present case $R = 1$ lb., $l = 48''$; therefore $P = 0·0208\,h$, where h is the height of the plane in inches, and P the power in pounds.

263. We compare the powers calculated by this formula with the actual observed values: the result is given in Table XIII.

TABLE XIII.—INCLINED PLANE.

Glass Plane 48″ long, truck 1 lb. in weight, friction counterpoised; formula $P = 0·0208 \times h''$.

Number of Experiment.	Height of plane.	Observed power in lbs.	P. Calculated power in lbs.	Difference of the observed and calculated powers.
1	2″	0·04	0·04	0·00
2	4″	0·08	0·08	0·00
3	6″	0·13	0·12	−0·01
4	8″	0·16	0·17	+0·01
5	10″	0·21	0·21	0·00
6	15″	0·31	0·31	0·00
7	20″	0·42	0·42	0·00
8	33″	0·71	0·69	−0·02

Thus for example, in experiment 6, where the height B E is 15″, it is observed that the power necessary to draw the truck is 0·31 lb. The truck is placed in the middle of the plane, and the power is adjusted so as to be sufficient to

IX.] INCLINED PLANE WITH FRICTION. 135

draw the truck to the top with certainty; the necessary power calculated by the formula is also 0·31 lbs., so that the theory is verified.

264. The fifth column of the table shows the difference between the observed and the calculated powers. The very slight differences, in no case exceeding the fiftieth part of a pound, may be referred to the inevitable errors of experiment.

THE INCLINED PLANE WITH FRICTION.

265. The friction of the truck upon the glass plate is always very small, and is shown to have but little variation at those inclinations of the plane which we used. But when the friction is large, we shall not be justified in neglecting its changes at different elevations, and we must adopt more rigorous methods. For this inquiry we shall use the pine plank and slide already described in Art. 117. We do not in this case attempt to diminish friction by the aid of wheels, and consequently it will be of considerable amount.

266. In another respect the experiments of Table XIII. are also in contrast with those now to be described. In the former the load was constant, while the elevation was changed. In the latter the elevation remains constant while a succession of different loads are tried. We shall find in this inquiry also that when the proper allowance has been made for friction, the theoretical law connecting the power and the load is fully verified.

267. The apparatus used is shown in Fig. 33; the plane, is, however, secured at one inclination, and the pulley c shown in Fig. 32 is adjusted to the apparatus, so that the rope from the pulley to the slide is parallel to the incline. The elevation of the plane in the position adopted is $17°·2$, so

that its length, base, and height are in the proportions of the numbers 1, 0·955, and 0·296. Weights ranging from 7 lbs. to 56 lbs. are placed upon the slide, and the power is found which, when the slide is started by the screw, will draw it steadily up the plane. The requisite power consists of two parts, that which is necessary to overcome gravity acting down the plane, and that which is necessary to overcome friction.

268. The forces are shown in Fig. 42. R G, the force of gravity, is resolved into R L and R M; R L is evidently the component acting down the plane, and R M the pressure against the plane; the triangle G L R is similar to A B C, hence if R be the load, the force R L acting down the plane must be 0·296 R, and the pressure upon the plane 0·955 R.

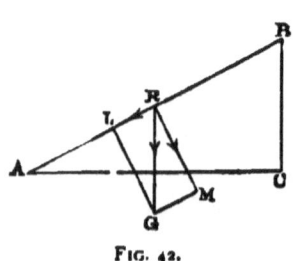

Fig. 42.

269. We shall first make a calculation with the ordinary law that the friction is proportional to the pressure. The pressure upon the plane A B, to which the friction is proportional, is not the weight of the load. The pressure is that component (R M) of the load which is perpendicular to the plane A B. When the weights do not extend beyond 56 lbs., the best value for the coefficient of friction is 0·288 (Art. 141): hence the amount of friction upon the plane is

$$0\cdot 288 \times 0\cdot 955\,R = 0\cdot 275\,R.$$

This force must be overcome in addition to 0·296 R (the component of gravity acting down the plane): hence the expression for the power is

$$0\cdot 275\,R + 0\cdot 296\,R = 0\cdot 571\,R.$$

IX.]   INCLINED PLANE WITH FRICTION.   137

270. The values of the observed powers compared with the powers calculated from the expression $0\cdot 571\,R$ are shown in Table XIV.

TABLE XIV.—INCLINED PLANE.

Smooth plane of pine $72''\times 11''$; angle of inclination $17°\cdot 2$; slide of pine, grain crosswise; slide started; formula $P=0\cdot 571\,R$.

Number of Experiment.	R. Total load on slide in lbs.	Power in lbs. which just draws up slide.	P. Calculated value of the power.	Difference of the observed and calculated powers.
1	7	4·6	4·0	−0·6
2	14	8·3	8·0	−0·3
3	21	12·3	12·0	−0·3
4	28	16·5	16·0	−0·5
5	35	20·0	20·0	0·0
6	42	24·2	24·0	−0·2
7	49	28·0	28·0	0·0
8	56	31·8	32·0	+0·2

271. Thus for example, in experiment 6, a load of 42 lbs. was raised by a force of 24·2 lbs., while the calculated value is 24·0 lbs.; the difference, 0·2 lbs., is shown in the last column.

272. The calculated values are found to agree tolerably well with the observed values, but the presence of the large differences in No. 1 and No. 4 leads us to inquire whether by employing the more accurate law of friction (Art. 141) a better result may not be obtained.

In Table VI. we have shown that the friction for weights not exceeding 56 lbs. is expressed by the formula $F = 0\cdot 9 + 0\cdot 266\times$ pressure, but the pressure is in this case $=0\cdot 955\,R$, and hence the friction is

$$0\cdot 9 + 0\cdot 254\,R.$$

To this must be added $0\cdot 296\,R$, the component of the force of gravity which must be overcome, and hence the total force necessary is

$$0.9 + 0.55 R.$$

The powers calculated from this expression are compared with those actually observed in Table XV.

TABLE XV.—INCLINED PLANE.

Smooth plane of pine 72" × 11"; angle of inclination 17°·2; slide of pine, grain crosswise; slide started; formula $P = 0.9 + 0.55 R$.

Number of Experiment.	R. Total load on slide in lbs.	Power in lbs. which just draws up slide.	P. Calculated value of the power.	Difference of the observed and calculated powers.
1	7	4·6	4·7	+0·1
2	14	8·3	8·6	+0·3
3	21	12·3	12·5	+0·2
4	28	16·5	16·3	−0·2
5	35	20·0	20·1	+0·1
6	42	24·2	24·0	−0·2
7	49	28·0	27·8	−0·2
8	56	31·8	31·7	−0·1

For example: in experiment 5, a load of 35 lbs. is found to be raised by a power of 20·0 lbs., while the calculated power is $0.9 + 0.55 \times 35 = 20.1$ lbs.

273. The calculated values of the powers are shown by this table to agree extremely well with the observed values, the greatest difference being only 0.3 lb. Hence there can be no doubt that the principles on which the formula has been calculated are correct. This table may therefore be regarded as verifying both the law of friction, and the rule laid down for the relation between the power and the load in the inclined plane.

274. The inclined plane is properly styled a mechanical power. For let the weight be 30 lbs., we calculate by the formula that 17·4 lbs. would be sufficient to raise it, so that,

notwithstanding the loss by friction, we have here a smaller force overcoming a larger one, which is the essential feature of a mechanical power. The mechanical efficiency is $30 \div 17\cdot4 = 1\cdot72$.

275. The velocity ratio in the inclined plane is the ratio of the distance through which the power moves to the height through which the weight is raised, that is $1 \div 0\cdot296 = 3\cdot38$. To raise 30 lbs. one foot, a force of $17\cdot4$ lbs. must therefore be exerted through $3\cdot38$ feet. The number of units of work expended is thus $17\cdot4 \times 3\cdot38 = 58\cdot8$. Of this 30 units, equivalent to 51 per cent., are utilized. The remaining $28\cdot8$ units, or 49 per cent., are absorbed by friction.

276. We have pointed out in Art. 222 that a machine in which less than half the energy is lost by friction will permit the load to run down when free : this is the case in the present instance ; hence the weight will run down the plane unless specially restrained. That it should do so agrees with Art. 147, for it was there shown that at about $13°\cdot4$, and still more at any greater inclination, the slide would descend when started.

THE SCREW.

277. The inclined plane as a mechanical power is often used in the form of a wedge or in the still more disguised form of a screw. A wedge is an inclined plane which is forced under the load; it is usually moved by means of a hammer, so that the efficiency of the wedge is augmented by the dynamical effect of a blow.

278. The screw is one of the most useful mechanical powers which we possess. Its form may be traced by wrapping a wedge-shaped piece of paper around a cylinder, and then cutting a groove in the cylinder along the spiral line indicated by the margin of the paper. Such a groove

is a screw. In order that the screw may be used it must revolve in a *nut* which is made from a hollow cylinder, the internal diameter of which is equal to that of the cylinder from which the screw is cut. The nut contains a spiral ridge, which fits into the corresponding thread in the screw; when the nut is turned round, it moves backwards or forwards according to the direction of the rotation. Large screws of the better class, such as those upon which we shall first make experiments, are always turned in a lathe, and are thus formed with extreme accuracy. Small screws are made in a simpler manner by means of dies and other contrivances.

279. A characteristic feature of a screw is the inclination of the thread to the axis. This is most conveniently described by the number of complete turns which the thread makes in a specified length of the screw, usually an inch. For example: a screw is said to have ten threads to the inch when it requires 10 revolutions of the nut in order to move it one inch. The shape of the thread itself varies with the purposes for which the screw is intended; the section may be square or triangular, or, as is generally the case in small screws, of a rounded form.

280. There is so much friction in the screw that experiments are necessary for the determination of the law connecting the power and the load.

281. We shall commence with an examination of the screw by the apparatus shown in Fig. 43.

The nut A is secured upon a stout frame; to the end of the screw hooks are attached, in order to receive the load, which in this apparatus does not exceed 224 lbs.; at the top of the screw is an arm E by which the screw is turned; to the end of the arm a rope is attached, which passing over a pulley D, carries a hook for receiving the power C.

THE SCREW.

282. We first apply the principle of work to this screw, and calculate the relation between the power and the load as it would be found if friction were absent. The diameter of the circle described by the end of the arm is 20"·5 ; its circumference is therefore 64"·4. The screw contains three threads in the inch, hence in order to raise the load 1" the

Fig. 43.

power moves $3 \times 64"·4 = 193"$ very nearly ; therefore the velocity ratio is 193, and were the screw capable of working without friction, 193 would represent the mechanical efficiency. In actually performing the experiments the arm E is placed at right angles to the rope leading to the pulley, and the power hook is weighted until, with a slight start, the arm is steadily drawn; but the power will only move the arm a few inches, for when the cord ceases to be perpendicular to the arm the power acts with diminished efficiency; consequently the load is only raised in each experiment through a small fraction of an inch, but quite sufficient for our purpose.

TABLE XVI.—THE SCREW.

Wrought iron screw, square thread, diameter 1"·25, with 3 threads to the inch, length of arm 10"·25; nut of cast iron, bearing surfaces oiled, velocity ratio 193, useful effect 36 per cent., mechanical efficiency 70; formula $P = 0·0143 R$.

Number of Experiment.	R. Load in lbs.	Observed power in lbs.	P. Calculated power of lbs.	Difference of the observed and calculated powers.
1	28	0·4	0·4	0·0
2	56	0·8	0·8	0·0
3	84	1·2	1·2	0·0
4	112	1·6	1·6	0·0
5	140	2·0	2·0	0·0
6	168	2·4	2·4	0·0
7	196	2·7	2·8	+0·1
8	224	3·3	3·2	−0·1

283. The results of the experiments are shown in Table XVI. If the motion had not been aided by a start the powers would have been greater. Thus in experiment 6, 2·4 lbs. is the power with a start, when without a start 3·2 lbs. was found to be necessary. The experiments have all been aided by a start, and the results recorded have been corrected for the friction of the pulley over which the rope passes: this correction is very small, in no case exceeding 0·2 lb. The fourth column contains the values of the powers computed by the formula $P = 0·0143 R$. This formula has been deduced from the observations in the

manner described in the Appendix. The fifth column proves that the experiments are truly represented by the formula: in each of the experiments 7 and 8, the difference between the calculated and observed values amounts to 0·1 lb., but this is quite inconsiderable in comparison with the weights we are employing.

284. In order to lift 100 lbs. the expression 0·0143 R shows that 1·43 lbs. would be necessary: hence the mechanical efficiency of the screw is 100 ÷ 1·43 = 70. Thus this screw is vastly more powerful than any of the pulley systems which we have discussed. A machine so capable, so compact, and so strong as the screw, is invaluable for innumerable purposes in the Arts, as well as in multitudes of appliances in daily use.

285. It is evident, however, that the distance through which the screw can raise a weight must be limited by the length cf the screw itself, and that in the *length of lift* the screw cannot compete with many of the other contrivances used in raising weights.

286. We have seen that the velocity ratio is 193; therefore, to raise 100 lbs. 1 foot, we find that 1·43 × 193 = 276 units of energy must be expended: of this only 100 units, or 36 per cent., is usefully employed; the rest being consumed in overcoming the friction of the screw. Thus nearly two-thirds of the energy applied to such a screw is wasted. Hence we find that friction does not permit the load to run down, since less than fifty per cent. of the applied energy is usefully employed (Art. 222). This is one of the valuable properties which the screw possesses.

287. We may contrast the screw with the pulley block (Art. 199). They are both powerful machines: the latter is bulky and economical of power, the former is compact and wasteful of power; the latter is adapted for raising

IX.] THE SCREW-JACK. 145

weights through considerable distances, and the former for exerting pressures through short distances.

THE SCREW-JACK.

288. The importance of the screw as a mechanical power justifies us in examining another of its useful forms, the screw-jack. This machine is used for exerting great pressures, such for example as starting a ship which is reluctant to be launched, or replacing a locomotive upon the line from which its wheels have slipped. These machines vary in form, as well as in the weights for which they are adapted; one of them is shown at D in Fig. 44, and a description of its details is given in Table XVII. We shall determine the powers to be applied to this machine for overcoming resistances not exceeding half a ton.

289. To employ weights so large as half a ton would be inconvenient if not actually impossible in the lecture-room, but the required pressures can be produced by means of a lever. In Fig. 44 is shown a stout wooden bar 16' long. It is prevented from bending by means of a chain; at E the lever is attached to a hinge, about which it turns freely; at A a tray is placed for the purpose of receiving weights. The screw-jack is 2' distant from E, consequently the bar is a lever of-the second order, and any weight placed in the tray exerts a pressure eightfold greater upon the top of the screw-jack. Thus each stone in the tray produces a pressure of 1 cwt. at the point D. The weight of the lever and the tray is counterpoised by the weight C, so that until the tray receives a load there is no pressure upon the top of the screw-jack, and thus we may omit the lever itself from consideration. The screw-jack is furnished with an arm D G; at the extremity G of this arm a rope is attached, which passes over a pulley and supports the power B.

290. The velocity ratio for this screw-jack with an arm of 33″, is found to be 414, by the method already described (Art. 283).

291. To determine its mechanical efficiency we must resort to experiment. The result is given in Table XVII.

TABLE XVII.—THE SCREW-JACK.

Wrought iron screw, square thread, diameter 2″, pitch 2 threads to the inch, arm 33″ ; nut brass, bearing surfaces oiled ; velocity ratio 414 ; useful effect, 28 per cent. ; mechanical efficiency 116; formula $P = 0.66 + 0.0075 \, R$.

Number of Experiment.	R. Load in lbs.	Observed power in lbs.	P. Calculated power in lbs.	Difference of the observed and calculated powers.
1	112	1·4	1·5	+0·1
2	224	2·2	2·3	+0·1
3	336	3·3	3·2	−0·1
4	448	4·1	4·0	−0·1
5	560	5·0	4·9	−0·1
6	672	5·7	5·7	0·0
7	784	6·5	6·5	0·0
8	896	7·4	7·4	0·0
9	1008	8·1	8·2	+0·1
10	1120	9·0	9·1	+0·1

292. It may be seen from the column of differences how closely the experiments are represented by the formula. The power which is required to raise a given weight, say 600 lbs., may be calculated by this formula; it is $0.66 + 0.0075 \times 600 = 5.16$. Hence the mechanical efficiency of the screw-jack is $600 \div 5.16 = 116$. Thus the screw is very powerful, increasing the force applied to it more than a hundredfold. In order to raise 600 lbs. one foot, a quantity of work represented by $5.16 \times 414 = 2136$ units must be ex-

pended; of this only 600, or 28 per cent., is utilized, so that nearly three-quarters of the energy applied is expended upon friction.

293. This screw does not let the load run down, since less than 50 per cent. of energy is utilised; to lower the weight the lever has actually to be pressed backwards.

294. The details of an experiment on this subject will be instructive, and afford a confirmation of the principles laid down. In experiment 10 we find that 9·0 lbs. suffice to raise 1,120 lbs.; now by moving the pulley to the other side of the lever, and placing the rope perpendicularly to the lever, I find that to produce motion the other way—that is, of course to lower the screw—a force of 3·4 lbs. must be applied. Hence, even with the assistance of the load, a force of 3·4 lbs. is necessary to overcome friction. This will enable us to determine the amount of friction in the same manner as we determined the friction in the pulley-block (Art. 207). Let x be the force usefully employed in raising, and y the force of friction, which acts equally in either direction against the production of motion; then to raise the load the power applied must be sufficient to overcome both x and y, and therefore we have $x+y=9\cdot0$. When the weight is to be lowered the force x of course aids in the lowering, but x alone is not sufficient to overcome the friction; it requires the addition of 3·4 lbs., and we have therefore $x+3\cdot4=y$, and hence $x=2\cdot8$, $y=6\cdot2$.

That is, 2·8 is the amount of force which with a frictionless screw would have been sufficient to raise half a ton. But in the frictionless screw the power is found by dividing the load by the velocity ratio. In this case $1120 \div 414 = 2\cdot7$, which is within 0·1 lb. of the value of x. The agreement of these results is satisfactory.

THE SCREW BOLT AND NUT.

295. One of the most useful applications of the screw is met with in the common bolt and nut, shown in Fig. 45. It consists of a wrought-iron rod with a head at one end and a screw on the other, upon which the nut works. Bolts in many different sizes and forms represent the stitches by which machines and frames are most readily united. There are several reasons why the bolt is so convenient. It draws the parts into close contact with tremendous force; it is itself so strong that the parts united practically form one piece. It can be adjusted quickly, and removed as readily. The same bolt by the use of washers can be applied to pieces of very different sizes. No skilled hand is required to use the simple tool that turns the nut. Adding to this that bolts are cheap and durable, we shall easily understand why they are so extensively used.

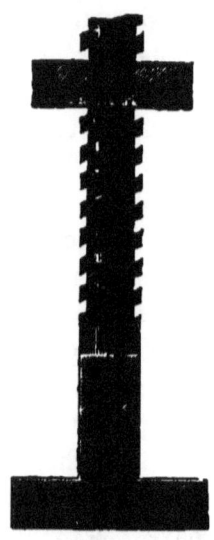

FIG. 45.

296. We must remark in conclusion that the bolt owes its utility to friction; screws of this kind do not overhaul, hence when the nut is screwed home it does not recoil. If it were not that more than half the power applied to a screw is consumed in friction, the bolt and the nut would either be rendered useless, or at least would require to be furnished with some complicated apparatus for preventing the motion of the nut.

LECTURE X.

THE WHEEL AND AXLE.

Introduction.—Experiments upon the Wheel and Axle.—Friction upon the Axle.—The Wheel and Barrel.—The Wheel and Pinion.—The Crane.—Conclusion.

INTRODUCTION.

297. THE mechanical powers discussed in these lectures may be grouped into two classes,—the first where ropes or chains are used, and the second where ropes or chains are absent. Belonging to that class in which ropes are not employed, we have the screw discussed in the last lecture, and the lever discussed in Lecture VIII.; while among those machines in which ropes or chains form an essential part of the apparatus, the pulley and the wheel and axle hold a prominent place. We have already examined several forms of the pulley, and we now proceed to the not less important subject of the wheel and axle.

298. Where great resistances have to be overcome, but where the distance through which the resistance must be urged is short, the lever or the screw is generally found to be the most appropriate means of increasing power. When,

however, the resistance has to be moved a considerable distance, the aid of the pulley, or the wheel and axle, or sometimes of both combined, is called in. The wheel and axle is the form of mechanical power which is generally used

Fig. 46.

when the distance is considerable through which a weight must be raised, or through which some resistance must be overcome.

299. The wheel and axle assumes very many forms corresponding to the various purposes to which it is applied.

X.] THE WHEEL AND AXLE. 151

The general form of the arrangement will be understood from Fig. 46. It consists of an iron axle B, mounted in bearings, so as to be capable of turning freely; to this axle a rope is fastened, and at the extremity of the rope is a weight D, which is gradually raised as the axle revolves. Attached to the axle, and turning with it, is a wheel A with hooks in its circumference, upon which lies a rope; one end of this rope is attached to the circumference of the wheel, and the other supports a weight E. This latter weight may be called the power, while the weight D suspended from the axle is the load. When the power is sufficiently large, E descends, making the wheel to revolve; the wheel causes the axle to revolve, and thus the rope is wound up and the load D is raised.

300. When compared with the differential pulley as a means of raising a weight, this arrangement appears rather bulky and otherwise inconvenient, but, as we shall presently learn, it is a far more economical means of applying energy. In its practical application, moreover, the arrangement is simplified in various ways, two of which may be mentioned.

301. The capstan is essentially a wheel and axle; the power is not in this case applied by means of a rope, but by direct pressure on the part of the men working it; nor is there actually a wheel employed, for the pressure is applied to what would be the extremities of the spokes of the wheel if a wheel existed.

302. In the ordinary winch, the power of the labourer is directly applied to the handle which moves round in the circumference of a circle.

303. There are innumerable other applications of the principle which are constantly met with, and which can be easily understood with a little attention. These we shall

not stop to describe, but we pass on at once to the important question of the relation between the power and the load.

EXPERIMENTS UPON THE WHEEL AND AXLE.

304. We shall commence a series of experiments upon the wheel A and axle B of Fig. 46. We shall first determine the velocity ratio, and then ascertain the mechanical efficiency by actual experiment. The wheel is of wood; it is about 30″ in diameter. The string to which the power is attached is coiled round a series of hooks, placed near the margin of the wheel; the effective circumference is thus a little less than the real circumference. I measure a single coil of the string and find the length to be 88″·5. This length, therefore, we shall adopt for the effective circumference of the wheel. The axle is 0″·75 in diameter, but its effective circumference is larger than the circle of which this length is the diameter.

305. The proper mode of finding the effective circumference of the axle in a case where the rope bears a considerable proportion to the axle is as follows. Attach a weight to the extremity of the rope sufficient to stretch it thoroughly. Make the wheel and axle revolve suppose 20 times, and measure the height through which the weight is lifted; then the one-twentieth part of that height is the effective circumference of the axle. By this means I find the circumference of the axle we are using to be 2″·87.

306. We can now ascertain the velocity ratio in this machine. When the wheel and axle have made one complete revolution the power has been lowered through a distance of 88″·5, and the load has been raised through 2″·87. This is evident because the wheel and axle are

attached together, and therefore each completes one revolution in the same time; hence the ratio of the distance which the power moves over to that through which the load is raised is $88''\cdot 5 \div 2''\cdot 87 = 31$ very nearly. We shall therefore suppose the velocity ratio to be 31. Thus this wheel and axle has a far higher velocity ratio than any of the systems of pulleys which we have been considering.

307. Were friction absent the velocity ratio of 31 would necessarily express the mechanical efficiency of this wheel and axle; owing to the presence of friction the real efficiency is less than this—how much less, we must ascertain by experiment. I attach a load of 56 lbs. to the hook which is borne by the rope descending from the axle: this load is shown at D in Fig. 46. I find that a power of 2·6 lbs. applied at E is just sufficient to raise D. We infer from this result that the mechanical efficiency of this machine is $56 \div 2\cdot 6 = 21\cdot 5$. I add a second 56 lb. weight to the load, and I find that a power of 5·0 lbs. raises the load of 112 lbs. The mechanical efficiency in this case is $112 \div 5 = 22\cdot 5$. We adopt the mean value 22. Hence the mechanical efficiency is reduced by friction from 31 to 22.

308. We may compute from this result the number of units of energy which are utilized out of every 100 units applied. Let us suppose a load of 100 lbs. is to be raised one foot; a force of $100 \div 22 = 4\cdot 6$ lbs. will suffice to raise this load. This force must be exerted through a space of 31', and consequently $31 \times 4\cdot 6 = 143$ units of energy must be expended; of this amount 100 units are usefully employed, and therefore the percentage of energy utilized is $100 \div 143 \times 100 = 70$. It follows that 30 per cent. of the applied energy is consumed in overcoming friction.

309. We can see the reason why the wheel and axle overhauls—that is, runs down of its own accord—when

allowed to do so; it is because less than half the applied energy is expended upon friction.

310. A series of experiments which have been carefully made with this wheel and axle are recorded in Table XVIII.

TABLE XVIII.—WHEEL AND AXLE.

Wheel of wood; axle of iron, in oiled brass bearings; weight of wheel and axle together, 16·5 lbs.; effective circumference of wheel, 88"·5; effective circumference of axle, 2"·87; velocity ratio, 31; mechanical efficiency, 22; useful effect, 70 per cent.; formula, $P = 0\cdot204 + 0\cdot0426\, R$.

Number of Experiment.	R. Load in lbs.	Observed power in lbs.	P. Calculated power in lbs.	Difference of the observed and calculated values.
1	28	1·4	1·4	0·0
2	42	2·0	2·0	0·0
3	56	2·6	2·6	0·0
4	70	3·2	3·2	0·0
5	84	3·7	3·8	+ 0·1
6	98	4·4	4·4	0·0
7	112	5·0	5·0	0·0

By the method of the Appendix a relation connecting the power and the load has been determined; it is expressed in the form—

$$P = 0\cdot204 + 0\cdot0426\, R.$$

311. Thus for example in experiment 5 a load of 84 lbs. was found to be raised by a power of 3·7 lbs. The value calculated by the formula is $0\cdot204 + 0\cdot0426 \times 84 = 3\cdot8$. The calculated value only differs from the observed value by 0·1 lb., which is shown in the fifth column. It will be seen from this column that the values calculated from the formula represents the experiments with fidelity.

312. We have deduced the relation between the power

FRICTION UPON THE AXLE.

and the load from the principle of energy, but we might have obtained it from the principle of the lever. The wheel and axle both revolve about the centre of the axle; we may therefore regard the centre as the fulcrum of a lever, and the points where the cords meet the wheel and axle as the points of application of the power and the load respectively.

313. By the principle of the lever of the first order (Art. 237), the power is to the load in the inverse proportion of the arms; in this case, therefore, the power is to the load in the inverse proportion of the radii of the wheel and the axle. But the circumferences of circles are in proportion to their radii, and therefore the power must be to the load as the circumference of the axle is to the circumference of the wheel.

314. This mode of arriving at the result is a little artificial; it is more natural to deduce the law directly from the principle of energy. In a mechanical power of any complexity it would be difficult to trace exactly the transmission of power from one part to the next, but the principle of energy evades this difficulty; no matter what be the mechanical arrangement, simple or complex, of few parts or of many, we have only to ascertain by trial how many feet the power must traverse in order to raise the load one foot; the number thus obtained is the theoretical efficiency of the machine.

FRICTION UPON THE AXLE.

315. In the wheel and axle upon which we have been experimenting, we have found that about 30 per cent. of the power is consumed by friction. We shall be able to ascertain to what this loss is due, and then in some degree to remove its cause. From the experiments of Art. 165

we learned that the friction of a small pulley was very much greater than that of a large pulley—in fact, the friction is inversely proportional to the diameter of the pulley. We infer from this that by winding the rope upon a barrel instead of upon the axle, the friction may be diminished.

FIG. 47.

316. We can examine experimentally the effect of friction on the axle by the apparatus of Fig. 47. B is a shaft $0''\cdot75$ diameter, about which a rope is coiled several times; the ends of this rope hang down freely, and to each of them hooks E, F are attached. This shaft revolves in brass

bearings, which are oiled. In order to investigate the amount of power lost by winding the rope upon an axle of this size, I shall place a certain weight—suppose 56 lbs.—upon one hook F, and then I shall ascertain what amount of power hung upon the other hook E will be sufficient to raise F. There is here no mechanical advantage, so that the excess of load which E must receive in order to raise F is the true measure of the friction.

317. I add on weights at E until the power reaches 85 lbs., when E descends. We thus see that to raise 56 lbs. an excess of 29 lbs. was necessary to overcome the friction. We may roughly enunciate the result by stating that to raise a load in this way, half as much again is required for the power. This law is verified by suspending 28 lbs. at F, when it is found that a power of 43 lbs. at E is required to lift it. Had the power been 42 lbs., it would have been exactly half as much again as the load.

318. Hence in raising F upon this axle, about one-third of the power which must be applied at the circumference of the axle is wasted. This experiment teaches us where the loss lies in the wheel and axle of Art. 304, and explains how it is that about a third of its efficiency is lost. 85 lbs. was only able to raise two-thirds of its own weight, owing to the friction; and hence we should expect to find, as we actually have found, that the power applied at the circumference of the wheel has an effect which is only two-thirds of its theoretical efficiency.

319. From this experiment we should infer that the proper mode of avoiding the loss by friction is to wind the rope upon a barrel of considerable diameter rather than upon the axle itself. I place upon a similar axle to that on which we have been already experimenting a barrel of about 15" circumference. I coil the rope two or three

times about the barrel, and let the ends hang down as before. I then attach to each end 56 lbs. weight, and I find that 10 lbs. added to either of the weights is sufficient to overcome friction, to make it descend, and raise the other weight. The apparatus is shown in Fig. 47. A is the barrel, C and D are the weights. In this arrangement 10 lbs. is sufficient to overcome the friction which required 29 lbs. when the rope was simply coiled around the axle. In other words, by the barrel the loss by friction is reduced to one-third of its amount.

THE WHEEL AND BARREL.

320. We next place the barrel upon the axis already experimented upon and shown in Fig. 46 at B. The circumference of the wheel is $88''\cdot 5$; the circumference of the barrel is $14''\cdot 9$. The proper mode of finding the circumference of the barrel is to suspend a weight from the rope, then raise this weight by making one revolution of the wheel, and the distance through which the weight is raised is the effective circumference of the barrel. The velocity ratio of the wheel and barrel is then found, by dividing $14\cdot 9$ into $88\cdot 5$, to be $5\cdot 94$.

321. The mechanical efficiency of this machine is determined by experiment. I suspend a weight of 56 lbs. from the hook, and apply power to the wheel. I find that $10\cdot 1$ lbs. is just sufficient to raise the load.

322. The mechanical efficiency is to be found by dividing $10\cdot 1$ into 56; the quotient thus obtained is $5\cdot 54$. The mechanical efficiency does not differ much from $5\cdot 94$, the velocity ratio; and consequently in this machine but little power is expended upon friction.

323. We can ascertain the loss by computing the percentage of applied energy which is utilized. Let us sup-

X.] THE WHEEL AND BARREL. 159

pose a weight of 100 lbs. has to be raised one foot: for this purpose a force of 100÷5·54 = 18·1 lbs. must be applied. This is evident from the definition of the mechanical efficiency; but since the load has to be raised one foot, it is clear from the meaning of the velocity ratio that the power must move over 5'·94: hence the number of units of work to be applied is to be measured by the product of 5·94 and 18·1, that is, by 107·5; in order therefore to accomplish 100 units of work 107·5 units of work must be applied. The percentage of energy usefully employed is 100÷107·5 × 100 = 93. This is far more than 70, which is the percentage utilized when the axle was used without the barrel (Art. 309).

324. A series of experiments made with care upon the wheel and barrel are recorded in Table XIX.

TABLE XIX.—THE WHEEL AND BARREL.

Wheel of wood, 88"·5 in circumference, on the same axle as a cast-iron barrel of 14"·9 circumference; axle is of wrought iron, 0"·75 in diameter, mounted in oiled brass bearings; power is applied to the circumference of the wheel, load raised by rope round barrel; velocity ratio, 5·94; mechanical efficiency, 5·54; useful effect, 93 per cent.; formula, $P = 0.5 + 0.169\,R$.

Number of Experiment.	R. Load in lbs.	Observed power in lbs	P. Calculated power in lbs.	Difference of the observed and calculated values.
1	14	2·7	2·9	+0·2
2	28	5·3	5·2	−0·1
3	42	7·7	7·6	−0·1
4	56	10·1	10·0	−0·1
5	70	12·4	12·4	0·0
6	84	14·7	14·7	0·0
7	98	17·1	17·1	0·0
8	112	19·4	19·5	+0·1

The formula which represents the experiments with the greatest amount of accuracy is $P = 0\cdot 5 + 0\cdot 169\, R$. This formula is compared with the experiments, and the column of differences shows that the calculated and the observed values agree very closely. The constant part 0·5 is partly due to the constant friction of the heavy barrel and wheel, and partly, it may be, to small irregularities which have prevented the centre of gravity of the whole mass from being strictly in the axle.

325. Though this machine is more economical of power than the wheel and axle of Art. 305, yet it is less powerful; in fact, the mechanical efficiency, 5·54, is only about one-fourth of that of the wheel and axle. It is therefore necessary to inquire whether we cannot devise some method by which to secure the advantages of but little friction, and at the same time have a large mechanical efficiency: this we shall proceed to investigate.

THE WHEEL AND PINION.

326. By means of what are called cog-wheels or toothed-wheels, we are enabled to combine two or more wheels and axles together, and thus greatly to increase the power which can be produced by a single wheel and axle. Toothed-wheels are used for a great variety of purposes in mechanics; we have already had some illustration of their use during these lectures (Fig. 30). The wheels which we shall employ are those often used in lathes and other small machines; they are what are called 10-pitch wheels,—that is to say, a wheel of this class contains ten times as many teeth in its circumference as there are inches in its diameter. I have here a wheel 20″ diameter, and consequently it has 200 teeth; here is another which is 2″·5 diameter, and which consequently contains 25 teeth. We shall mount these

wheels upon two parallel shafts, so that they gear one into the other in the manner shown in Fig 46: F is the large wheel containing 200 teeth, and G the pinion of 25 teeth. The axles are $0''\cdot 75$ diameter; around each of them a rope is wound, by which a hook is suspended.

327. A small weight at K is sufficient to raise a much larger weight on the other shaft; but before experimenting on the mechanical efficiency of this arrangement, we shall as usual calculate the velocity ratio. The wheel contains eight times as many teeth as the pinion; it is therefore evident that when the wheel has made one revolution, the pinion will have made eight revolutions, and conversely the pinion must turn round eight times to turn the wheel round once: hence the power which is turning the pinion round must be lowered through eight times the circumference of the axle, while the load is raised through a length equal to one circumference of the axle. We thus find the velocity ratio of the machine to be 8.

328. We determine the mechanical efficiency by trial. Attaching a load of 56 lbs. to the axle of the large wheel, it is observed that a power of $13\cdot 7$ lbs. at K will raise it; the mechanical efficiency of the machine is therefore about $4\cdot 1$, which is almost exactly half the velocity ratio. We note that the load will only just run down when the power is removed; from this we might have inferred, by Art. 222, that nearly half the power is expended on friction, and that therefore the mechanical efficiency is about half the velocity ratio. The actual percentage of energy that is utilised with this particular load is 51. If we suspend 112 lbs. from the load hook, 26 lbs. is just enough to raise it; the mechanical efficiency that would be deduced from this result is $112 \div 26 = 4\cdot 3$, which is slightly in excess of the amount obtained by the former experiment. It is often found to be

M

a property of the mechanical powers, that as the load increases the mechanical efficiency slightly improves.

329. In Table XX. will be found a record of experiments upon the relation between the power and the load with the wheel and pinion; the table will sufficiently explain itself, after the description of similar tables already given (Arts. 310, 324).

TABLE XX.—THE WHEEL AND PINION.

Wheel (10-pitch), 200 teeth; pinion, 25 teeth; axles equal, effective circumference of each being $2''\cdot 87$; oiled brass bearings; velocity ratio, 8; mechanical efficiency, $4\cdot 1$; useful effect, 51 per cent.; formula, $P = 2\cdot 46 + 0\cdot 21\ R$.

Number of Experiment.	R. Load in lbs.	Observed power in lbs.	P. Calculated power in lbs.	Difference of the observed and calculated powers.
1	14	5·4	5·4	0·0
2	28	8·7	8·3	−0·4
3	42	11·0	11·3	+0·3
4	56	13·7	14·2	+0·5
5	70	17·5	17·2	−0·3
6	84	20·0	20·1	+0·1
7	98	23·0	23·0	0·0
8	112	26·0	26·0	0·0

330. The large amount of friction present in this contrivance is the consequence of winding the rope directly upon the axle instead of upon a barrel, as already pointed out in Art. 319. We might place barrels upon these axles and demonstrate the truth of this statement; but we need not delay to do so, as we use the barrel in the machines which we shall next describe.

THE CRANE.

331. We have already explained (Art. 38) the construction of the lifting crane, so far as its framework is concerned. We now examine the mechanism by which the load is raised. We shall employ for this purpose the model which is repre-

sented in Fig. 48. The jib is supported by a wooden bar as a tie, and the crane is steadied by means of the weights placed at H: some such counterpoise is necessary, for otherwise the machine would tumble over when a load is suspended from the hook.

332. The load is supported by a rope or chain which passes over the pulley E and thence to the barrel D, upon which it is to be wound. This barrel receives its motion from a large wheel A, which contains 200 teeth.

The wheel A is turned by the pinion B which contains 25 teeth. In the actual use of the crane, the axle which carries this pinion would be turned round by means of a handle; but for the purpose of experiments upon the relation of the power to the load, the handle would be inconvenient, and therefore we have placed upon the axle of the pinion a wheel C containing a groove in its circumference. Around this groove a string is wrapped, so that when a weight G is suspended from the string it will cause the wheel to revolve. This weight G will constitute the power by which the load may be raised.

333. Let us compute the velocity ratio of this machine before commencing experiments upon its mechanical efficiency. The effective circumference of the barrel D is found by trial to be $14''\cdot 9$. Since there are 200 teeth on A and 25 on B, it follows that the pinion B must revolve eight times to produce one revolution of the barrel. Hence the wheel C at the circumference of which the power is applied must also revolve eight times for one revolution of the barrel. The effective circumference of C is $43''$; the power must therefore have been applied through $8 \times 43'' = 344''$, in order to raise the load $15''\cdot 9$. The velocity ratio is $344 \div 14\cdot 9 = 23$ very nearly. We can easily verify this value of the velocity ratio by actually raising the load $1'$, when it appears that the

FIG. 48.

LECT. X.] THE CRANE. 165

number of revolutions of the wheel B is such that the power must have moved 23'.

334. The mechanical efficiency is to be found as usual by trial. 56 lbs. placed at F is raised by 3·1 lbs. at G; hence the mechanical efficiency deduced from this experiment is $56 \div 3 \cdot 1 = 18$. The percentage of useful effect is easily shown to be 78 by the method of Art. 323. Here, then, we have a machine possessing very considerable efficiency, and being at the same time economical of energy.

TABLE XXI.—THE CRANE.

Circumference of wheel to which the power is applied, 43"; train of wheels, 25 ÷ 200; circumference of drum on which rope is wound, 14"·9; velocity ratio, 23; mechanical efficiency, 18; useful effect, 78 per cent.; formula, $P = 0\cdot0556\,R$.

Number of Experiment.	R. Load in lbs.	Observed power in lbs.	P. Calculated power in lbs.	Difference of the observed and calculated values.
1	14	0·9	0·8	−0·1
2	28	1·6	1·6	0·0
3	42	2·4	2·3	−0·1
4	56	3·1	3·1	0·0
5	70	3·8	3·9	+0·1
6	84	4·5	4·7	+0·2
7	98	5·3	5·5	+0·2
8	112	6·2	6·2	+0·0

335. A series of experiments made with this crane is recorded in Table XXI., and a comparison of the calculated and observed values will show that the formula $P = 0\cdot0556\,R$ represents the experiments with considerable accuracy.

336. It may be noticed that in this formula the term independent of R, which we frequently meet with in the expression of the relation between the power and the load, is absent. The probable explanation is to be found in the fact that some minute irregularity in the form of the

barrel or of the wheel has been constantly acting like a small weight in favour of the power. In each experiment the motion is always started from the same position of the wheels, and hence any irregularity will be constantly acting in favour of the power or against it ; here the former appears to have happened. In other cases doubtless the latter has occurred; the difference is, however, of extremely small amount. The friction of the machine itself when without a load is another source for the production of the constant term; it has happened in the present case that this friction has been almost exactly balanced by the accidental influence referred to.

337. In cranes it is usual to provide means of adding a second train of wheels, when the load is very heavy. In another model we applied the power to an axle with a pinion of 25 teeth, gearing into a wheel of 200 teeth ; on the axle of the wheel with 200 teeth is a pinion of 30 teeth, which gears into a wheel of 180 teeth ; the barrel is on the axle of the last wheel. A series of experiments with this crane is shown in Table XXII.

TABLE XXII.—THE CRANE FOR HEAVY LOADS.

Circumference of wheel to which power is applied, 43"; train of wheels, 25 ÷ 200 × 30 ÷ 180; circumference of drum on which rope is wound, 14"·9 ; velocity ratio, 137 ; mechanical efficiency, 87 ; useful effect, 63 per cent. ; formula, $P = 0.185 + 0.00782\, R$.

Number of Experiment.	R. Load in lbs.	Observed power in lbs.	P. Calculated power in lbs.	Difference of the observed and calculated values.
1	14	0·30	0·29	−0·01
2	28	0·40	0·40	0·00
3	42	0·50	0·51	+0·01
4	56	0·60	0·62	+0·02
5	70	0·75	0·73	−0·02
6	84	0·85	0 84	−0·01
7	98	0·95	0·95	0·00
8	112	1·05	1·06	+0·01

X.] THE CRANE. 167

The velocity ratio is now 137, and the mechanical efficiency is 87; one man could therefore raise a ton with ease by applying a power of 26 lbs. to a crane of this kind.

CONCLUSION.

338. It will be useful to contrast the wheel and axle on which we have experimented (Art. 304) with the differential pulley (Art. 209). The velocity ratio of the former machine is nearly double that of the latter, and its mechanical efficiency is nearly four times as great. Less than half the applied power is wasted in the wheel and axle, while more than half is wasted in the differential pulley. This makes the wheel and axle both a more powerful machine, and a more economical machine than the differential pulley. On the other hand, the greater compactness of the latter, its facility of application, and the practical conveniences arising from the property of not allowing the load to run down, do often more than compensate for the superior mechanical advantage of the wheel and axle.

339. We may also contrast the wheel and axle with the screw (Art. 277). The screw is remarkable among the mechanical powers for its very high velocity ratio, and its excessive friction. Thus we have seen in Art. 291 how the velocity ratio of a screw-jack with an arm attached amounted to 414, while its mechanical efficiency was little more than one-fourth as great. No *single* wheel and axle could conveniently be made to give a mechanical efficiency of 116; but from Art. 337 we could easily design a *combination* of wheels and axles to yield an efficiency of quite this amount. The friction in the wheel and axle is very much less than in the screw, and consequently energy is saved by the use of the former machine.

340. In practice, however, it generally happens that economy of energy does not weigh much in the selection of a mechanical power for any purpose, as there are always other considerations of greater consequence.

341. For example, let us take the case of a lifting crane employed in loading or unloading a vessel, and inquire why it is that a train of wheels is generally used for the purpose of producing the requisite power. The answer is simple, the train of wheels is convenient, for by their aid any length of chain can be wound upon the barrel; whereas if a screw were used, we should require a screw as long as the greatest height of lift. This screw would be inconvenient, and indeed impracticable, and the additional circumstance that a train of wheels is more economical of energy than a screw has no influence in the matter.

342. On the other hand, suppose that a very heavy load has to be overcome for a short distance, as for example in starting a ship launch, a screw-jack is evidently the proper machine to employ; it is easily applied, and has a high mechanical efficiency. The want of economy of energy is of no consequence in such an operation.

LECTURE XI.

THE MECHANICAL PROPERTIES OF TIMBER.

Introduction.—The General Properties of Timber.—Resistance to Extension.—Resistance to Compression.—Condition of a Beam strained by a Transverse Force.

INTRODUCTION

343. IN the lectures on the mechanical powers which have been just completed, we have seen how great weights may be raised or other large resistances overcome. We are now to consider the important subject of the application of mechanical principles to *structures*. These are fixtures, while machines are adapted for motion; a roof or a bridge is a structure, but a crane or a screw-jack is a machine. Structures are employed for supporting weights, and the mechanical powers give the means of raising them.

344. A structure has to support both its own weight and also any load that is to be placed upon it. Thus a railway bridge must at all times sustain what is called the permanent load, and frequently, of course, the weight of one or more trains. The problem which the engineer solves is to design a bridge which shall be sufficiently strong, and, at the same

time, economical; his skill is shown by the manner in which he can attain these two ends in the same structure.

345. In the four lectures of the course which will be devoted to this subject it will only be possible to give a slight sketch, and therefore but few details can be introduced. An extended account of the properties of different materials used in structures would be beyond our scope, but there are some general principles relating to the strength of materials which may be discussed. Timber, as a building material, has, in modern times, been replaced to a great extent by iron in large structures, but timber is more capable than iron of being experimented upon in the lecture room. The elementary laws which we shall demonstrate with reference to the strength of timber, are also, substantially the same as the corresponding laws for the strength of iron or any other material. Hence we shall commence the study of structures by two lectures on timber. The laws which we shall prove experimentally will afterwards be applied to a few simple cases of bridges and other actual structures.

THE GENERAL PROPERTIES OF TIMBER.

346. The uses of timber in the arts are as various as its qualities. Some woods are useful for their beauty, and others for their strength or durability under different circumstances. We shall only employ "pine" in our experimental inquiries. This wood is selected because it is so well known and so much used. A knowledge of the properties of pine would probably be more useful than a knowledge of the properties of any other wood, and at the same time it must be remembered that the laws which we shall establish by means of slips of pine may be generally applied.

347. A transverse section of a tree shows a number of rings, each of which represents the growth of wood in one

year. The age of the tree may sometimes be approximately found by counting the number of distinguishable rings. The outer rings are the newer portions of the wood.

348. When a tree is felled it contains a large quantity of sap, which must be allowed to evaporate before the wood is fit for use. With this object the timber is stored in suitable yards for two or more years according to the purposes for which it is intended; sometimes the process of seasoning, as it is called, is hastened by other means. Wood, when seasoning, contracts; hence blocks of timber are often found split from the circumference to the centre, for the outer rings, being newer and containing more sap, contract more than the inner rings. For the same reason a plank is found to warp when the wood is not thoroughly seasoned. The side of the plank which was farthest from the centre of the tree contracts more than the other side, and becomes concave. This can be easily verified by looking at the edge of the plank, for we there see the rings of which it is composed.

349. Timber may be softened by steaming. I have here a rod of pine, $24'' \times 0'''.5 \times 0'''.5$, and here a second rod cut from the same piece and of the same size, which has been exposed to steam of boiling water for more than an hour: securing these at one end to a firm stand, I bend them down together, and you see that after the dry rod has broken the steamed rod can be bent much farther before it gives way. This property of wood is utilized in shaping the timbers of wooden ships. We shall be able to understand the action of steam if we reflect that wood is composed of a number of fibres ranged side by side and united together. A rope is composed of a number of fibres laid together and twisted, but the fibres are not coherent as they are in wood. Hence we find that a rod of wood is stiff, while

a rope is flexible. The steam finds its way into the interstices between the fibres of the wood; it softens their connections, and increases the pliability of the fibres themselves, and thus, the operation of steaming tends to soften a piece of timber and render it tractable.

350. The structure of wood is exhibited by the following simple experiment :—Here are two pieces of pine, each 9" × 1" × 1". One of them I easily snap across with a blow, while my blows are unable to break the other. The difference is merely that one of these pieces is cut against the grain, while the other is with it. In the first case I have only to separate the connection between the fibres, which is quite easy. In the other case I would have to tear asunder the fibres themselves, which is vastly more difficult. To a certain extent the grained structure is also found in wrought iron, but the contrast between the strength of iron with the grain and against the grain is not so marked as it is in wood.

RESISTANCE TO EXTENSION.

351. It will be necessary to explain a little more definitely what is meant by the strength of timber. We may conceive a rod to be broken in three different ways. In the first place the rod may be taken by a force at each end and torn asunder by pulling, as a thread may be broken. To do this requires very great power, and the strength of the material with reference to such a mode of destroying it is called its resistance to extension. In the second place, it may be broken by longitudinal pressure at each end, as a pillar may be crushed by the superincumbent weight being too large; the strength that relates to this form of force is called resistance to compression: finally, the rod may be broken by a force applied transversely. The strength of pine with reference to these different applications of force will be

XI.] RESISTANCE TO EXTENSION. 173

considered successively. The rods that are to be used have been cut from the same piece of timber, which has been selected on account of its straightness of grain and freedom from knots. They are of different rectangular sections,

FIG. 49.

$1'' \times 0''\cdot 5$ and $0''\cdot 5 \times 0''\cdot 5$ being generally used, but sometimes $1 \times 1''$ is employed.

352. With reference to the strength of timber in its capacity to resist extension, we can do but little in the lecture room. I have here a pine rod A B, of dimensions $48'' \times 0''\cdot 5 \times 0''\cdot 5$, Fig. 49. Each end of this rod is firmly secured

between two cheeks of iron, which are bolted together: the rod is suspended by its upper extremity from the hook of the epicyloidal pulley-block (Art. 213), which is itself supported by a tripod; hooks are attached to the lower end of the rod for carrying the weights. By placing 3 cwt. on these hooks and pulling the hand chain of the pulley-block, I find that I can raise the weight safely, and therefore the rod will resist at all events a tension of 3 cwt. From experiments which have been made on the subject, it is ascertained that about a ton would be necessary to tear such a rod asunder; hence we see that pine is enormously strong in resisting a force of extension. The tensile strength of the rod does not depend upon its length, but upon the area of the cross section. That of the rod we have used is one-fourth of a square inch, and the breaking weight of a rod one square inch in section is about four tons.

353. A rod of any material generally elongates to some extent under the action of a suspended weight; and we shall ascertain whether this occurs perceptibly in wood. Before the rod was strained I had marked two points upon it exactly 2 feet apart. When the rod supports 3 cwt. I find that the distance between the two points has not appreciably altered, though by more delicate measurement I have no doubt we should find that the distance had elongated to an insignificant extent.

354. Let us contrast the resistance of a rod of timber to extension with the effect upon a rope under the same circumstances. I have here a rope about $0''.25$ diameter; it is suspended from a point, and bears a 14 lb. weight in order to be completely stretched. I mark points upon the rope $2'$ apart. I now change the stone weight for a weight of 1 cwt., and on measurement I find that the two points which before were $2'$ apart, are now $2' 2''$; thus the

rope has stretched at the rate of an inch per foot for a strain of 1 cwt., while the timber did not stretch perceptibly for a strain of 3 cwt.

355. We have already explained in Art. 37 the meaning of the word "tie." The material suitable for a tie should be capable of offering great resistance, not only to actual rupture by tension, but even to appreciable elongation. These qualities we have found to be possessed by wood. They are, however, possessed in a much higher degree by wrought iron, which possesses other advantages in durability and facility of attachment.

RESISTANCE TO COMPRESSION.

356. We proceed to examine into the capability of timber to resist forces of longitudinal compression, either as a pillar or in any other form of "strut," such for instance, as the jib of the crane represented in Fig. 17. The use of timber as a strut depends in a great degree upon the coherence of the fibres to each other, as well as upon their actual rigidity. The action of timber in resisting forces of compression is thus very different from its action when resisting forces of extension; we can examine, by actual experiment, the strength of timber under the former conditions, as the weights which it will be necessary to employ are within the capabilities of our lecture-room apparatus.

357. The apparatus is shown in Fig. 50. It consists of a lever of the second order, 10' long, the mechanical advantage of which is threefold; the resistance of the pillar D E to crushing is the load to be overcome, and the power consists of weights, to receive which the tray B is used; every pound placed in the tray produces a compressive force of 3 lbs. on the pillar at D. The fulcrum is at A and guides at G. The lever and the tray would somewhat complicate our

calculations unless their weights were counterpoised. A cord attached to the extremity of the lever passes over a pulley F; at the other end of this cord, sufficient weights C are attached to neutralize the weight of the apparatus. In fact, the lever and tray now swing as if they had no weight, and we may therefore leave them out of consideration. The pillar to be experimented upon is fitted at its lower

FIG. 50.

end E into a hole in a cast-iron bracket: this bracket can be adjusted so as to take in pieces of different lengths; the upper end of the pillar passes through a hole in a second piece of cast-iron, which is bolted to the lever: thus our little experimental column is secured at each end, and the risk of slipping is avoided. The stands are heavily weighted to secure the stability of the arrangement.

358. The first experiment we shall make with this

apparatus is upon a pine rod 40″ long and 0″·5 square; the lower bracket is so placed that the lever is horizontal when just resting upon the top of the rod. Weights placed in the tray produce a pressure three times as great down the rod, the effect of which will first be to bend the rod, and, when the deflection has reached a certain amount, to break it across. I place 28 lbs. in the tray: this produces a pressure of 84 lbs. upon the rod, but the rod still remains perfectly straight, so that it bears this pressure easily. When the pressure is increased to 96 lbs. a very slight amount of deflection may be seen. When the strain reaches 114 lbs. the rod begins to bend into a curved form, though the deflection of the middle of the rod from its original position is still less than 0″·25. Gradually augmenting the pressure, I find that when it reaches 132 lbs. the deviation has reached 0″·5; and finally, when 48 lbs. is placed in the tray, that is, when the rod is subjected to 144 lbs., it breaks across the middle. Hence we see that this rod sustained a load of 96 lbs. without sensibly bending, but that fracture ensued when the load was increased about half as much again. Another experiment with a similar rod gave a slightly less value (132 lbs.) for the breaking load. If I add these results together, and divide the sum by 2, I find 138 lbs. as the mean value of the breaking load, and this is a sufficiently exact determination.

359. Let us next try the resistance of a shorter rod of the same section. I place a piece of pine 20″ long and 0″·5 square in the apparatus, firmly securing each end as in the former case. The lower bracket is adjusted so as to make the lever horizontal; the counterpoise, of course, remains the same, and weights are placed in the tray as before. No deflection is noticed when the rod supports

126 lbs.; a very slight amount of bending is noticeable with 186 lbs.; with 228 lbs., the amount by which the centre of the rod has deviated laterally from its original position is about $0''\cdot 2$; and finally, when the load reaches 294 lbs., the rod breaks. Fracture first occurs in the middle, but is immediately followed by other fractures near where the ends of the rod are secured.

360. Hence the breaking load of a rod 20″ long is more than double the breaking load of a rod of 40″ long the same section; from this we learn that the sections being equal, short pillars are stronger than long pillars. It has been ascertained by experiment that the strength of a square pillar to resist compression is proportional to the square of its sectional area. Hence a rod of pine, 40″ long and 1″ square, having four times the section of the rod of the same length we have experimented on, would be sixteen times as strong, and consequently its breaking weight would amount to nearly a ton. The strength of a rod used as a *tie* depends only on its section, while the strength of a rod used as a *strut* depends on its length as well as on its section.

CONDITION OF A BEAM STRAINED BY A TRANSVERSE FORCE.

361. We next come to the important practical subject of the strength of timber when supporting a transverse strain; that is, when used as a beam. The nature of a transverse strain may be understood from Fig. 51, which represents a small beam, strained by a load at its centre. Fig. 52 shows two supports 40″ apart, across which a rod of pine 48″ × 1″ × 1″ is laid; at the middle of this rod a hook is placed, from which a tray for the reception of weights is suspended. A rod thus supported, and bearing weights, is

said to be strained transversely. A rafter of a roof, the flooring of a room, a gangway from the wharf to a ship, many forms of bridge, and innumerable other examples, might be given of beams strained in this manner. To this important subject we shall devote the remainder of this lecture and the whole of the next.

362. The first point to be noticed is the deflection of the beam from which a weight is suspended. The beam is at first horizontal; but as the weight in the tray is

Fig. 51.

augmented, the beam gradually curves downwards until, when the weight reaches a certain amount, the beam breaks across in the middle and the tray falls.

For convenience in recording the experiments the tray chain and hooks have been adjusted to weigh exactly 14 lbs. (Fig. 52). A B is a cord which is kept stretched by the little weights D: this cord gives a rough measure of the deflection of the beam from its horizontal position when strained by a load in the tray. In order to observe the deflection

180   EXPERIMENTAL MECHANICS.   [LECT.

accurately an instrument is used called the cathetometer (G). It consists of a small telescope, always directed horizontally, though capable of being moved up and down a vertical triangular pillar; on one of the sides of the pillar a scale is engraved, so that the height of the telescope in any

FIG. 52.

position can be accurately determined. The cathetometer is levelled by means of the screws H H, so that the triangular pillar on which the telescope slides is accurately vertical: the dotted line shows the direction of the visual ray when the centre C of the beam is seen by the observer through the telescope.

XI.] TRANSVERSE STRAIN. 181

Inside the telescope and at its focus a line of spider's web is fixed horizontally; on the bar to be observed, and near its middle point C, a cross of two fine lines is marked. The tray being removed, the beam becomes horizontal; the telescope of the cathetometer is then directed towards the beam, so that the lines marked upon it can be seen distinctly. By means of a screw the telescope may be raised or lowered until the spider's web inside the telescope is observed to pass through the image of the intersection of the lines. The scale then indicates precisely how high the telescope is on the pillar.

363. While I look through the telescope my assistant suspends the tray from the beam. Instantly I see the cross descend in the field of view. I lower the telescope until the spider's web again passes through the image of the intersection of the lines, and then by looking at the scale I see that the telescope has been moved down $0''\cdot 19$, that is, about one-fifth of an inch: this is, therefore, the distance by which the cross lines on the beam, and therefore the centre of the beam itself, must have descended. Indeed, even a simpler apparatus would be competent to measure the amount of deflection with some degree of precision. By placing successively one stone after another upon the tray, the beam is seen to deflect more and more, until even without the telescope you see the beam has deviated from the horizontal.

364. By carefully observing with the telescope, and measuring in the way already described, the deflections shown in Table XXIII. were determined. The scale along the vertical pillar was read after the spider's web had been adjusted for each increase in the weight. The movement from the original position is recorded as the deflection for each load.

TABLE XXIII.—DEFLECTION OF A BEAM.

A rod of pine 48″ × 1″ × 1″; resting freely on supports 40″ apart; and laden in the middle.

Number of Experiment.	Magnitude of load.	Deflection.
1	14	0″·19
2	28	0″·37
3	42	0″·55
4	56	0″·74
5	70	0″·94
6	84	1″·13
7	98	1″·35
8	112	1″·61
9	126	1″·95
10	140	2″·37

365. The first column records the number of the experiment. The second represents the load, and the third contains the corresponding deflections. It will be seen that up to 98 lbs. the deflection is about 0″·2 for every stone weight, but afterwards the deflection increases more rapidly. When the weight reaches 140 lbs. the deflection at first indicated is 2″·37; but gradually the cross lines are seen to descend in the field of the telescope, showing that the beam is yielding and finally it breaks across. This experiment teaches us that a beam is at first deflected by an amount proportional to the weight it supports; but that when two-thirds of the breaking weight is reached, the beam is deflected more rapidly.

366. It is a question of the utmost importance to ascertain the greatest load a beam can sustain without injury to its strength. This subject is to be studied by examining the effect of different deflections upon the fibres of a beam. A beam is always deflected whatever be the load it supports;

thus by looking through the telescope of the cathetometer I can detect an increase of deflection when a single pound is placed in the tray: hence whenever a beam is loaded we must have some deflection. An experiment will show what amount of deflection may be experienced without producing any permanently injurious effect.

367. A pine rod 40″ × 1 × 1″ is freely supported at each end, the distances between the supports being 38″, and the tray is suspended from its middle point. A fine pair of cross lines is marked upon the beam, and the telescope of the cathetometer is adjusted so that the spider's line exactly passes through the image of the intersection. 14 lbs. being placed in the tray, the cross is seen to descend; the weight being removed, the cross returns precisely to its original position with reference to the spider's line: hence, after this amount of deflection, the beam has clearly returned to its initial condition, and is evidently just as good as it was before. The tray next received 56 lbs.; the beam was, of course, considerably deflected, but when the weight was removed the cross again returned,—at all events, to within 0″·01 of where the spider's line was left to indicate its former position. We may consider that the beam is in this case also restored to its original condition, even though it has borne a strain which, including the tray, amounted to 70 lbs. But when the beam has been made to carry 84 lbs. for a few seconds, the cross does not completely return on the removal of the load from the tray, but it shows that the beam has now received a permanent deflection of 0″·03. This is still more apparent after the beam has carried 98 lbs., for when this load is removed the centre of the beam is permanently deflected by 0″·13. Here, then, we may infer that the fibres of the beam are beginning to be strained beyond their powers of resistance, and this is

verified when we find that with 28 additional pounds in the tray a collapse ensues.

368. Reasoning from this experiment, we might infer that the elasticity of a beam is not affected by a weight which is less than half that which would break it, and that, therefore, it may bear without injury a weight not exceeding this amount. As, however, in our experiments the weight was only applied once, and then but for a short time, we cannot be sure that a longer-continued or more frequent application of the same load might not prove injurious; hence, to be on the safe side, we assume that one-third of the breaking weight of a beam is the greatest load it should be made to bear in any structure. In many cases it is found desirable to make the beam much stronger than this ratio would indicate.

369. We next consider the condition of the fibres of a beam when strained by a transverse force. It is evident that since the fracture commences at the lower surface of the beam, the fibres there must be in a state of tension, while those at the concave upper surface of the beam are compressed together. This condition of the fibres may be proved by the following experiment.

370. I take two pine rods, each $48'' \times 1'' \times 1''$, perfectly similar in all respects, cut from the same piece of timber, and therefore probably of very nearly identical strength. With a fine tenon saw I cut each of the rods half through at its middle point. I now place one of these beams on the supports $40''$ apart, with the cut side of the beam upwards. I suspend from it the tray, which I gradually load with weights until the beam breaks, which it does when the total weight is 81 lbs.

If I were to place the second beam on the same supports with the cut upwards, then there can be no doubt that it would require as nearly as possible the same weight to break it. I place it, however, with the cut downwards, I suspend the tray, and find that the beam breaks with a load of 31 lbs. This is less than half the weight that would have been required if the cut had been upwards.

371. What is the cause of this difference? The fibres being compressed together on the upper surface, a cut has no tendency to open there; and if the cut could be made with an extremely fine saw, so as to remove but little material, the beam would be substantially the same as if it had not been tampered with. On the other hand, the fibres at the lower surface are in a state of tension; therefore when the cut is below it yawns open, and the beam is greatly weakened. It is, in fact, no stronger than a beam of $48'' \times 0''{\cdot}5 \times 1''$, placed with its shortest dimension vertical. If we remember that an entire beam of the same size required about 140 lbs. to break it (Art. 366), we see that the strength of a beam is reduced to one-fourth by being cut half-way through and having the cut underneath.

372. We may learn from this the practical consequence that the sounder side of a beam should always be placed downwards. Any flaw on the lower surface will seriously weaken the beam: so that the most knotty face of the wood should certainly be placed uppermost. If a portion of the actual substance of a beam be removed—for example, if a notch be cut out of it—this will be almost equally injurious on either side of the beam.

373. We may illustrate the condition of the upper surface of the beam by a further experiment. I make two cuts $0''{\cdot}5$ deep in the middle of a pine rod $48'' \times 1'' \times 1''$. These cuts are $0''{\cdot}5$ apart, and slightly inclined; the piece between them being removed, a wedge is shaped to fit tightly into the space; the wedge is long enough to project a little on one

side. If the wedge be uppermost when the beam is placed on the supports, the beam will be in the same condition as if it had two fine cuts on the upper surface. I now load the beam with the tray in the usual manner, and I find it to bear 70 lbs. securely. On examining the beam, which has curved down considerably, I find that the wedge is held in very tightly by the pressure of the fibres upon it, but, by a sharp tap at the end, I knock out the wedge, and instantly the load of 70 lbs. breaks the beam; the reason is simple—the piece being removed, there is no longer any resistance to the compressive strain of the upper fibres, and consequently the beam gives way.

374. The collapse of a beam by a transverse strain commences by fracture of the fibres on the lower surface, followed by a rupture of all fibres up to a considerable depth. Here we see that by a transverse force the fibres in a beam of $48'' \times 1'' \times 1''$ have been broken by a strain of 140 lbs. (Art. 366); but we have already stated (Art. 353) that to tear such a rod across by a direct pull at each end a force of about four tons is necessary. The breaking strain of the fibres must be a certain definite quantity, yet we find that to overcome it in one way four tons is necessary, while by another mode of applying the strain 140 lbs. is sufficient.

375. To explain this discrepancy we may refer to the experiment of Art. 28, wherein a piece of string was broken by the transverse pull of a piece of thread in illustration of the fact that one force may be resolved into two others, each of them very much greater than itself. A similar resolution of force occurs in the transverse deflection of the beam, and the force of 140 lbs. is changed into two other forces, each of them enormously greater and sufficiently strong to rupture the fibres. We need not suppose that

the force thus developed is so great as four tons, because that is the amount required to tear across a square inch of fibres simultaneously, whereas in the transverse fracture the fibres appear to be broken row after row; the fracture is thus only gradual, nor does it extend through the entire depth of the beam.

376. We shall conclude this lecture with one more remark, on the condition of a beam when strained by a transverse force. We have seen that the fibres on the upper surface are compressed, while those on the lower surface are extended; but what is the condition of the fibres in the interior? There can be no doubt that the following is the state of the case :—The fibres immediately beneath the upper surface are in compression; at a greater depth the amount of compression diminishes until at the middle of the beam the fibres are in their natural condition; on approaching the lower surface the fibres commence to be strained in extension, and the amount of the extension gradually increases until it reaches a maximum at the lower surface.

LECTURE XII.

THE STRENGTH OF A BEAM.

A Beam free at the Ends and loaded in the Middle.—A Beam uniformly loaded.—A Beam loaded in the Middle, whose Ends are secured.— A Beam supported at one end and loaded at the other.

A BEAM FREE AT THE ENDS AND LOADED IN THE MIDDLE.

377. In the preceding lecture we have examined some general circumstances in connection with the condition of a beam acted on by a transverse force; we proceed in the present to inquire more particularly into the strength under these conditions. We shall, as before, use for our experiments rods of pine only, as we wish rather to illustrate the general laws than to determine the strength of different materials. The strength of a beam depends upon its length, breadth, and thickness; we must endeavour to distinguish the effects of each of these elements on the capacity of the beam to sustain its load.

We shall only employ beams of rectangular section; this being generally the form in which beams of wood are used. Beams of iron, when large, are usually not rectangular, as the material can be more effectively disposed in sections of a different form. It is important to distinguish between the *stiffness* of a beam in its capacity to resist flexure, and the *strength* of a beam in its capacity to resist fracture. Thus the stiffest beam which can be made from the cylindrical trunk of a tree 1' in diameter is 6" broad and 10"·5 deep, while the strongest beam is 7" broad and 9"·75 deep. We are now discussing the strength (not the stiffness) of beams.

378. We shall commence the inquiry by making a number of experiments: these we shall record in a table, and then we shall endeavour to see what we can learn from an examination of this table. I have here ten pieces of pine, of lengths varying from 1' to 4', and of three different sections, viz. 1" × 1", 1" × 0"·5, and 0"·5 × 0"·5. I have arranged four different stands, on which we can break these pieces: on the first stand the distance between the points of support is 40", and on the other stands the distances are 30," 20", and 10" respectively; the pieces being 4', 3', 2', and 1' long, will just be conveniently held on the supports.

379. The mode of breaking is as follows:—The beam being laid upon the supports, an S hook is placed at its middle point, and from this S hook the tray is suspended. Weights are then carefully added to the tray until the beam breaks; the load in the tray, together with the weight of the tray, is recorded in the table as the breaking load.

380. In order to guard as much as possible against error, I have here another set of ten pieces of pine, duplicates of the former. I shall also break these; and whenever I find any difference between the breaking loads of two similar beams, I shall record in the table the mean between the two loads. The results are shown in Table XXIV.

TABLE XXIV.—STRENGTH OF A BEAM.

Slips of pine (cut from the same piece) supported freely at each end; the length recorded is the distance between the points of support; the load is suspended from the centre of the beam, and gradually increased until the beam breaks;

$$\text{Formula, } P = 6080 \; \frac{\text{area of section} \times \text{depth}}{\text{span}}$$

No. of Experiment.	Dimensions.			Mean of the observations of the breaking load in lbs.	P. Calculated breaking load in lbs.	Difference of the observed and calculated values.
	Span.	Breadth.	Depth.			
1	40"·0	1"·0	1"·0	152	152	0·0
2	40"·0	0"·5	1"·0	77	76	−1·0
3	40"·0	1"·0	0"·5	38	38	0·0
4	40"·0	0"·5	0"·5	19	19	0·0
5	30"·0	1"·0	0"·5	59	51	−8·0
6	30"·0	0"·5	0"·5	25	25	0·0
7	20"·0	1"·0	0"·5	74	76	+2·0
8	20"·0	0"·5	0"·5	36	38	+2·0
9	10"·0	1"·0	0"·5	154	152	−2·0
10	10"·0	0"·5	0"·5	68	76	+8·0

381. In the first column is a series of figures for convenience of reference. The next three columns are occupied with the dimensions of the beams. By span is meant the distance between the points of support; the real length is of course greater; the depth is that dimension of the beam which is vertical. The fifth column gives the mean of two observations of the breaking load. Thus for example, in experiment No. 5 the two beams used were each 36" × 1" × 0"·5, they were placed on points of support 30" distant, so the span recorded is 30": one of the beams

was broken by a load of 58 lbs., and the second by a load of 60 lbs. ; the mean between the two, 59 lbs., is recorded as the mean breaking load. In this manner the column of breaking loads has been found. The meaning of the two last columns of the table will be explained presently.

382. We shall endeavour to elicit from these observations the laws which connect the breaking load with the span, breadth, and depth of the beam.

383. Let us first examine the effect of the span ; for this purpose we bring together the observations upon beams of the same section, but of different spans. Sections of $0''·5 \times 0''·5$ will be convenient for this purpose ; Nos. 4, 6, 8, and 10 are experiments upon beams of this section. Let us first compare 4 and 8. Here we have two beams of the same section, and the span of one (40") is double that of the other (20"). When we examine the breaking weights we find that they are 19 lbs. and 36 lbs. ; the former of these numbers is rather more than half of the latter. In fact, had the breaking load of 40" been $\frac{3}{4}$ lb. less, 18·25 lbs., and had that of 20" been $\frac{1}{2}$ lb. more, 36·5 lbs., one of the breaking loads would have been exactly half the other.

384. We must not look for perfect numerical accuracy in these experiments ; we must only expect to meet with approximation, because the laws for which we are in search are in reality only approximate laws. Wood itself is variable in quality, even when cut from the same piece : parts near the circumference are different in strength from those nearer the centre; in a young tree they are generally weaker, and in an old tree generally stronger. Minute differences in the grain, greater or less perfectness in the seasoning, these are also among the circumstances which prevent one piece of timber from being identical with another. We shall, however, generally find that the

effect of these differences is small, but occasionally this is not the case, and in trying many experiments upon the breaking of timber, discrepancies occasionally appear for which it is difficult to account.

385. But you will find, I think, that, making reasonable allowances for such difficulties as do occur, the laws on the whole represent the experiments very closely.

386. We shall, then, assume that the breaking weight of a bar of 40″ is half that of a bar of 20″ of the same section, and we ask, Is this generally true? is it true that the breaking weight is inversely proportional to the span? In order to test this hypothesis, we can calculate the breaking weight of a bar of 30″ (No. 6), and then compare the result with the observed value; if the supposition be true, the breaking weight should be given by the proportion—

$$30'' : 40'' :: 19 : \text{Answer.}$$

The answer is 25·3 lbs.; on reference to the table we find 25 lbs. to be the observed value, hence our hypothesis is verified for this bar.

387. Let us test the law also for the 10″ bar, No. 10—

$$10'' : 40'' :: 19 : \text{Answer.}$$

The answer in this case is 76, whereas the observed value is 68, or 8 lbs. less; this does not agree very well with the theory, but still the difference, though 8 lbs., is only about 11 or 12 per cent. of the whole, and we shall still retain the law, for certainly there is no other that can express the result as well.

388. But the table will supply another verification. In experiment No. 3 a 40″ bar, 1″ broad, and 0″·5 deep, broke with 38 lbs.; and in experiment No. 7 a 20″ bar of

the same section broke with 74 lbs.; but this is so nearly double the breaking weight of the 40" bar, as to be an additional illustration of the law, that *for a given section the breaking load varies inversely as the span.*

389. We next inquire as to the effect of the *breadth* of the beam upon its strength? For this purpose we compare experiments Nos. 3 and 4: we there find that a bar 40"×1"×0"·5 is broken by a load of 38 lbs., while a bar just half the breadth is broken by 19 lbs. We might have anticipated this result, for it is evident that the bar of No. 3 must have the same strength as two bars similar to that of No. 4 placed side by side.

390. This view is confirmed by a comparison of Nos. 7 and 8, where we find that a 20" bar takes twice the load to break it that is required for a bar of half its breadth. The law is not quite so well verified by Nos. 5 and 6, for half the breaking weight of No. 5, namely 29·5 lbs., is more than 25, the observed breaking weight of No. 6: a similar remark may be made about Nos. 9 and 10.

391. Supposing we had a beam of 40" span, 2" broad, and 0"·5 deep, we can easily see that it is equivalent to two bars like that of No. 3 placed side by side; and we infer generally that the strength of a bar is proportional to its breadth; or to speak-more definitely, *if two beams have the same span and depth, the ratio of their breaking loads is the same as the ratio of their breadths.*

392. We next examine the effect of the *depth* of a beam upon its strength. In experimenting upon a beam placed edgewise, a precaution must be observed, which would not be necessary if the same beam were to be broken flatwise. When the load is suspended, the beam, if merely laid edgewise on the supports, would almost certainly turn over; it is therefore necessary to place its extremities in recesses in

O

the supports, which will obviate the possibility of this occurrence; at the same time the ends must not be prevented from bending upwards, for we are at present discussing a beam free at each end, and the case where the ends are not free will be subsequently considered.

393. Let us first compare together experiments Nos. 2 and 3; here we have two bars of the same dimensions, the section in each being $1''{\cdot}0 \times 0'''{\cdot}5$, but the first bar is broken edgewise, and the second flatwise. The first breaks with 77 lbs., and the second with 38 lbs.; hence the same bar is twice as strong placed edgewise as flatwise when one dimension of the section is twice as great as the other. We may generalize this law, and assert that *the strength of a rectangular beam broken edgewise is to the strength of a beam of like span and section broken flatwise, as the greater dimension of the section is to the lesser dimension.*

394. The strength of a beam $40'' \times 0'''{\cdot}5 \times 1''$ is four times as great as the strength of $40'' \times 0'''{\cdot}5 \times 0'''{\cdot}5$, though the quantity of wood is only twice as great in one as in the other. In general we may state that if a beam were bisected by a longitudinal cut, the strength of the beam would be halved when the cut was horizontal, and unaltered when the cut was vertical; thus, for example, two beams of experiment No. 4, placed one on the top of the other, would break with about 40 lbs., whereas if the same rods were in one piece, the breaking load would be nearly 80 lbs.

395. This may be illustrated in a different manner. I have here two beams of $40'' \times 1'' \times 0'''{\cdot}5$ superposed; they form one beam, equivalent to that of No. 1 in bulk, but I find that they break with 80 lbs., thus showing that the two are only twice as strong as one.

396. I take two similar bars, and, instead of laying them loosely one on the other, I unite them tightly with iron

clamps like those represented in Fig. 56. I now find that the bars thus fastened together require 104 lbs. for fracture. We can readily understand this increase of strength. As soon as the bars begin to bend under the action of the weight, the surfaces which are in contact move slightly one upon the other in order to accommodate themselves to the change of form. By clamping I greatly impede this motion hence the beams deflect less, and require a greater load before they collapse; the case is therefore to some extent approximated to the state of things when the two rods form one solid piece, in which case a load of 152 lbs. would be required to produce fracture.

397. We shall be able by a little consideration to understand the reason why a bar is stronger edgewise than flatwise. Suppose I try to break a bar across my knee by pulling the ends held one in each hand, what is it that resists the breaking? It is chiefly the tenacity of the fibres on the convex surface of the bar. If the bar be edgewise, these fibres are further away from my knee and therefore resist with a greater moment than when the bar is flatwise: nor is the case different when the bar is supported at each end, and the load placed in the centre; for then the reactions of the supports correspond to the forces with which I pulled the ends of the bar.

398. We can now calculate the strength of any rectangular beam of pine:

Let us suppose it to be 12′ long, 5″ broad, and 7″ deep. This is five times as strong as a beam 1″ broad and 7″ deep for we may conceive the original beam to consist of 5 of these beams placed side by side (Art 391); the beam 1″ broad and 7″ deep, is 7 times as strong as a beam 7″ broad, 1″ deep (Art. 393). Hence the original beam must be 35 times as strong as a beam 7″ broad, 1″ deep; but the beam 7″ broad and 1″ deep is seven times stronger than a beam the section of which is 1″×1″, hence the original beam is 245 times as strong as a beam 12′ long and 1″×1″ in section; of which we can calculate the strength, by Art. 388, from the proportion—

$$144'' : 40'' :: 152 : \text{Answer.}$$

The answer is 42·2 lbs., and thus the breaking load of the original beam is about 10,300 lbs.

399. It will be useful to deduce the *general* expression for the breaking load of a beam l'' span, b'' broad, and d'' deep, supported freely at the ends and laden in the centre.

Let us suppose a bar l'' long, and 1″ × 1″ in section. The breaking load is found by the proportion—

$$l : 40 :: 152 : \text{Answer};$$

and the result obtained is $\dfrac{6080}{l}$. A beam which is d'' broad, l'' span, and 1″ deep, would be just as strong as d of the beams $l'' \times 1'' \times 1$ placed side by side; of which the collective strength would be—

$$\frac{6080}{l} \times d.$$

If such a beam, instead of resting flatwise, were placed edgewise, its strength would be increased in the ratio of its depth to its breadth—that is, it would be increased d-fold—and would therefore amount to

$$\frac{6080}{l} \times d^2.$$

We thus learn the strength of a beam 1″ broad, d'' deep, and l'' span. The strength of b of these beams placed side by side, would be the same as the strength of one beam b'' broad, d'' deep, and l'' span, and thus we finally obtain

$$\frac{6080}{l} \times d^2 \times b.$$

Since $b\ d$ is the area of the section, we can express this result conveniently by saying that the breaking load in lbs. of a rectangular pine beam is equal to

$$6080 \times \frac{\text{area of section} \times \text{depth}}{\text{span}};$$

the depth and span being expressed in inches linear measure, and the section in square inches.

400. In order to test this formula, we have calculated from it the breaking loads of all the ten beams given in Table XXIV. and the results are given in the sixth column. The difference between the amount calculated and the observed mean breaking weight is shown in the last column.

401. Thus, for example, in experiment No. 7 the span is 20″, breadth, 1″, depth 0″·5; the formula gives, since the area is 0″·5,

$$P = 6080 \frac{0\cdot 5 \times 0\cdot 5}{20} = 76.$$

This agrees sufficiently with 74 lbs., the mean of two observed values.

402. Except in experiments Nos. 5 and 10, the differences are very small, and even in these two cases the differences are not sufficient to make us doubt that we have discovered the correct expression for the load generally sufficient to produce fracture.

403. We have already pointed out that a beam begins to sustain permanent injury when it is subjected to a load greater than half that which would break it (Art. 368), and we may infer that it is not in general prudent to load a beam

which is part of a permanent structure with more than about a third of a fourth of the breaking weight. Hence if we wanted to calculate a fair working load in lbs. for a beam of pine, we might obtain it from the formula.

$$1500 \times \frac{\text{area of section} \times \text{depth}}{\text{span}}.$$

Probably a smaller coefficient than 1500 would often be used by the cautious builder, especially when the beam was liable to sudden blows or shocks. The coefficient obtained from small selected rods such as we have used would also be greater than that found from large beams in which imperfections are inevitable.

404. Had we adopted any other kind of wood we should have found a similar formula for the breaking weight, but with a different numerical coefficient. For example, had the beams been made of oak the number 6080 must be replaced by a larger figure.

A BEAM UNIFORMLY LOADED.

405. We have up to the present only considered the case where the load is suspended from the centre of the beam. But in the actual employment of beams the load is not generally applied in this manner. See in the rafters which support a roof how every inch in the entire length has its burden of slates to bear. The beams which support a warehouse floor have to carry their load in whatever manner the goods are disposed: sometimes, as for example in a grain-store, the pressure will be tolerably uniform along the beams, while if the weights be irregularly scattered on the floor, there will be corresponding inequalities in the mode in which the loads are distributed over the beams. It will therefore be useful for us to examine the strength of a beam when its load is applied otherwise than at the centre.

BEAM UNIFORMLY LOADED.

406. We shall employ, in the first place, a beam 40″ span, 0″·5 broad, and 1″ deep ; and we shall break it by applying a load simultaneously at two points, as may be most conveniently done by the contrivance shown in the diagram, Fig. 53. A B is the beam resting on two supports; C and D are the points of trisection of the span; from whence loops descend, which carry an iron bar P Q; at the centre R of which a weight W is suspended. The load is thus divided equally between the two points C and D, and we may regard A B as a beam loaded at its two points of trisection.

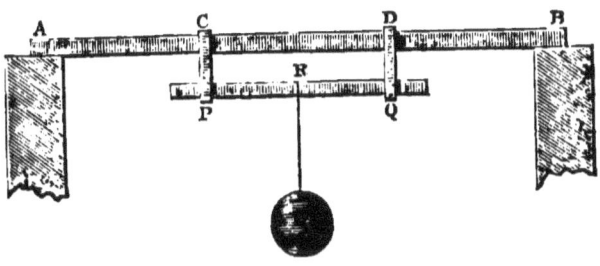

FIG. 53.

The tray and weights are employed which we have used in the apparatus represented in Fig. 58.

407. We proceed to break this beam. Adding weights to the tray, we see that it yields with 117 lbs., and cracks across between C and D. On reference to Table XXIV. we find from experiment No. 2 that a similar bar was broken by 77 lbs. at the centre; now $\frac{3}{2} \times 77 = 115·5$; hence we may state with sufficient approximation that the bar is half as strong again when the load is suspended from the two points of trisection as it is when suspended from the centre. It is remarkable that in breaking the beam in this manner the fracture is equally likely to occur at any point between C and D.

408. A beam *uniformly* loaded requires twice as much load to break it as would be sufficient if the load were merely suspended from the centre.* The mode of applying a load uniformly is shown in Fig. 54.

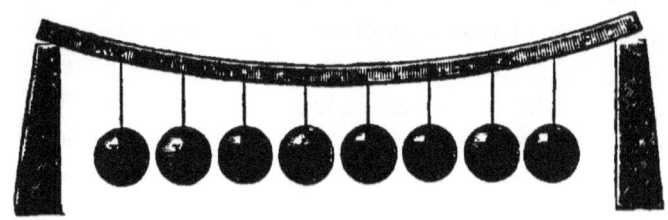

Fig. 54.

In an experiment actually tried, a beam 40″ × 0″·5 × 1″ placed edgewise was found to support ten 14 lb. weights ranged as in the figure; one or two stone more would, however, doubtless produce fracture.

409. We infer from these considerations that beams loaded in the manner in which they are usually employed are considerably stronger than would be indicated by the results in Table XXIV.

EFFECT OF SECURING THE ENDS OF A BEAM UPON ITS STRENGTH.

410. It has been noticed during the experiments that when the weights are suspended from a beam and the beam begins to deflect, the ends curve upwards from the supports. This bending of the ends is for example shown in Fig. 54. If we restrain the ends of the beam from bending up in this manner, we shall add very considerably to its strength. This we can do by clamping them down to the supports.

411. Let us experiment upon a beam 40″ × 1″ × 1″. We clamp each of the ends and then break the beam by a weight

XII.] BEAM SECURED AT ONE END. 201

suspended from the centre. It requires 238 lbs. to accomplish fracture. This is a little more than half as much again as 152 lbs., which we find from Table XXIV. was the weight required to break this bar when its ends were free. Calculation shows that the strength of a beam may be even doubled when the ends are kept horizontal by more perfect methods than we have used.

412. When the beam gives way under these circumstances, there is not only a fracture in the centre, but each of the halves are also found to be broken across near the points of support; the necessity for three fractures instead of one explains the increase of strength obtained by restraining the ends to the horizontal direction.

413. In structures the beams are generally more or less secured at each end, and are therefore more capable of bearing resistance than would be indicated by Table XXIV. From the consideration of Arts. 408 and 411, we can infer that a beam secured at each end and uniformly loaded would require three or four times as much load to break it as would be sufficient if the ends were free and if the load were applied at the centre.

BEAMS SECURED AT ONE END AND LOADED AT THE OTHER.

414. A beam, one end of which is firmly imbedded in masonry or otherwise secured, is occasionally called upon to support a weight suspended from its extremity. Such a beam is shown in Fig. 55.

In the case we shall examine, A B is a pine beam of dimensions $20'' \times 0''\cdot5 \times 0'''5$, and we find that, when W reaches 10 lbs., the beam breaks. In experiment No. 8, Table XXIV., a similar beam required 36 lbs.; hence we see that the beam is broken in the manner of Fig. 55, by about one-fourth of the load which would have been required if the beam had been supported at each end and laden in the centre.

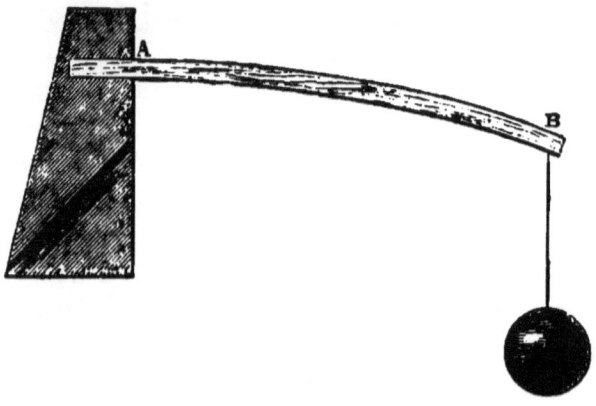

FIG. 55.

We shall presently have occasion to apply some of the results obtained by the experiments made in the lecture now terminated.

LECTURE XIII.

THE PRINCIPLES OF FRAMEWORK.

Introduction.—Weight sustained by Tie and Strut.—Bridge with Two Struts.—Bridge with Four Struts.—Bridge with Two Ties.—Simple Form of Trussed Bridge.

INTRODUCTION.

415. IN this lecture and the next we shall experiment upon some of the arts of construction. We shall employ slips of pine $0''\cdot5 \times 0''\cdot5$ in section for the purpose of making models of simple framework: these slips can be attached to each other by means of the small clamps about $3''$ long, shown in Fig. 56, and the general appearance of the models thus produced may be seen from Figs. 58 and 62.

FIG. 56.

416. The following experiment shows the tenacity with which these clamps hold. Two slips of pine, each $12'' \times 0''\cdot5 \times 0''\cdot5$, are clamped together, so that they overlap about $2''$, thus forming a length of $22''$: this composite rod is raised by a pulley-block as in Fig. 49, while a load of 2 cwt. is suspended from it. Thus the clamped rods bear a direct

tension of 2 cwt. The efficiency of the clamps depends principally upon friction, aided doubtless by a slight crushing of the wood, which brings the surfaces into perfect contact.

417. These slips of pine united by the clamps are possessed of strength quite sufficient for the experiments now to be described. Models thus constructed have the great advantage of being erected, varied or pulled down, with the utmost facility.

We have learned that the compressive strength, and, still more, the tensile strength of timber, is much greater than its transverse strength. This principle is largely used in the arts of construction. We endeavour by means of suitable combinations to turn transverse forces into forces of tension or compression, and thus strengthen our constructions. We shall illustrate the mode of doing so by simple forms of framework.

WEIGHT SUSTAINED BY TIE AND STRUT.

418. We begin with the study of a very simple contrivance, represented in Fig. 57.

A B is a rod of pine 20″ long. In the diagram it is represented, for simplicity, imbedded at the end A in the support. In reality, however, it is clamped to the support, and the same remark may be made about some other diagrams used in this lecture. Were A B unsupported except at its end A, it would of course break when a weight of 10 lbs. was suspended at B, as we have already found in Art. 414.

419. We must ascertain whether the transverse force on A B cannot be changed into forces of tension and compression. The tie B C is attached by means of clamps; A B is sustained by this tie; it cannot bend downwards under the action of the weight w, because we should then

XIII.] WEIGHT SUSTAINED BY TIE AND STRUT. 205

require to have on the same base and on the same side of it two triangles having their conterminous sides equal, but this we know from Euclid (I. 7) is impossible. Hence B is supported, and we find that 112 lbs. may be safely suspended, so that the strength is enormously increased. In fact the transverse force is changed into a compressive force or thrust down A B, and a tensile force on B C.

420. The actual magnitudes of these can be computed. Draw the parallelogram C D E B; if B D represent

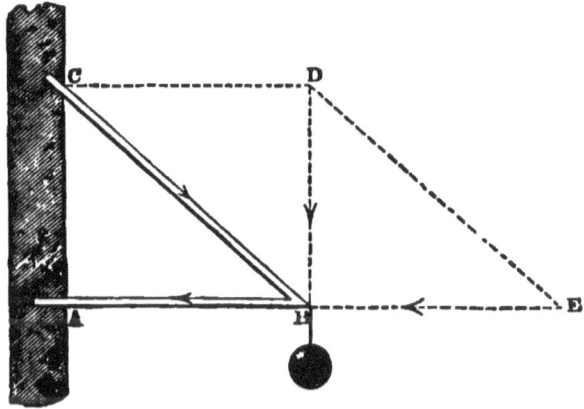

FIG. 57.

the weight W, it may be resolved into two forces,—one, B C, a force of extension on the tie; the other, B E, a compressive force on A B, which is therefore a strut. Hence the forces are proportional to the sides of the triangle, A B C. In the present case

$$A B = 20'', A C = 18'', B C = 27'';$$

therefore, when W is 112 lbs., we calculate that the force on A B is 124 lbs., and on C B 168 lbs. A B would require about 300 lbs. to crush it, and C B about 2,000 lbs. to tear it asunder, consequently the tie and strut can support 1 cwt. with ease. If, however, W were increased to about 270 lbs., the force on A B would become too great, and fracture would arise from the collapse of this strut.

421. When a structure is loaded up to the breaking point of one part, it is proper for economy that all the other parts should be so designed that they shall be as near as possible to their breaking points. In fact, since nothing is stronger than its weakest part, any additional strength which the remaining parts may possess adds no strength to the whole, and is only so much material wasted. Hence our structure would be just as strong, and would be more properly designed if the section of B C were reduced to one-fifth, for the tie would then break when the tension upon it amounted to 400 lbs. When W is 270 lbs. the compression on A B is 300 lbs., and the tension on B C is 405 lbs., so that both tie and strut attain their breaking loads together. The principle of duly apportioning the strength of each piece to the load it has to carry, involves the essence of sound engineering. In that greatest of mechanical feats, the construction of a mighty railway bridge across a wide span, attention to this principle is of vital importance. Such a bridge has to bear the occasional load of a passing train, but it has always to support the far greater load of the bridge materials. There is thus every inducement to make the weight of each part of the bridge as light as may be consistent with safety.

A BRIDGE WITH TWO STRUTS.

422. We shall next examine the structure of a type of bridge, shown in Fig. 58.

It consists of two beams, A B, 4' long, placed parallel to

XIII.] A BRIDGE WITH TWO STRUTS. 207

each other at a distance of 3″·5, and supported at each end; they are firmly clamped to the supports, and a roadway of short pieces is laid upon them. At the points of

Fig. 58.

trisection of the beams C, D, struts C F and D E are clamped, their lower ends being supported by the framework: these struts are 2′ long, and there are two of them supporting each of the beams. The tray G is attached by a chain to a stout piece of wood, which rests upon the roadway at the centre of the bridge.

423. We shall first determine the strength of this bridge by actual experiment, and then we shall endeavour to explain the results in accordance with mechanical principles. We can observe the deflection of the bridge by the cathetometer in the manner already described (Art. 362). By this means we shall ascertain whether the load has permanently injured the elasticity of the structure (Art. 367). We begin by testing the deflection when a load is distributed uniformly, as the weights are disposed in the case of Fig. 62. A cross is marked upon one of the beams, and is viewed in the cathetometer. We arrange 11 stone weights along the bridge, and the cathetometer shows that the deflection is only $0''\cdot09$: the elasticity of the bridge remains unaltered, for when the weights are removed the cross on the beam returns to its original position; hence the bridge is well able to bear this load.

424. We remove the row of weights from the bridge and suspend the tray from the roadway. I take my place at the cathetometer to note the deflection, while my assistant places weights H H on the tray. 1 cwt. being the load, I see that the deflection amounts to $0''\cdot2$; with 2 cwt. the deflection reaches $0\cdot43''$; and the bridge breaks with 238 lbs.

425. Let us endeavour to calculate the additional strength which the struts have imparted to the bridge. By Table XXIV. we see that a rod $40'' \times 0'''\cdot5 \times 0''\cdot5$ is broken by a load of 19 lbs.: hence the beams of the bridge would have been broken by a load of 38 lbs. if their ends had been free. As, however, the ends of the beams had been clamped down, we learn from Art. 411 that a double load would be necessary.

A BRIDGE WITH TWO STRUTS.

We may, however, be confident that about 80 lbs. would have broken the unsupported bridge. The strength is, therefore, increased threefold by the struts, for a load of 238 lbs. was required to produce fracture.

426. We might have anticipated this result, because the points C and D being supported by the struts may be considered as almost fixed points; in fact, we see that C cannot descend, because the triangle A C F is unalterable, and for a similar reason D remains fixed: the beam breaks between C and D, and the force required must therefore be sufficient to break a beam supported at the points C and D, whose ends are secured. But C D is one-third of A B, and we have already seen that the strength of a beam is inversely as its length (Art. 388); hence the force required to break the beam when supported by the struts is three times as large as would have been necessary to break the unsupported beam. Thus the strength of the bridge is explained.

427. As a load of 238 lbs. applied near the centre is necessary to break this bridge, it follows from the principle of Art. 408 that a load of about double this amount must be placed uniformly on the roadway before it succumbs; we can, therefore, understand how a load of 11 stone was easily borne (Art. 423) without permanent injury to the elasticity of the structure. If we take the factor of safety as 3, we see that a bridge of the form we have been considering may carry, as its ordinary working load, a far greater weight than would have crushed it if unsupported by the struts and with free ends.

428. The strength of the bridge in Fig. 58 is greater in some parts than in others. At the points C and D a maximum load could be borne; the weakest places on the bridge are in the middle points of the segments A C, D C,

and D B. The load applied by the tray was principally borne at the middle of D C, but owing to the piece of wood which sustained the chain being about 18" long, the load was to some extent distributed.

The thrust upon the struts is not so easy to calculate accurately. That down C F for example must be less than if the part C D were removed, and half the load were suspended from C. The force in this case can be determined by principles already explained (Art. 420).

A BRIDGE WITH FOUR STRUTS.

429. The same principles that we have employed in the construction of the bridge of Fig. 58 may be extended further, as shown in the diagram of Fig. 59.

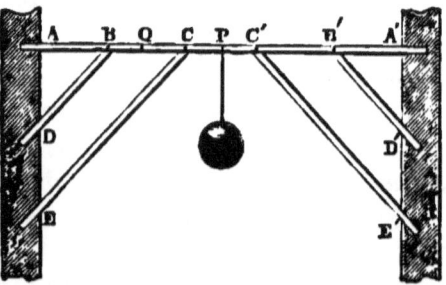

FIG. 59.

We have here two horizontal rods, 48" × 0"·5 × 0"·5, each end being secured to the supports; one of these rods is shown in the figure. It is divided into five equal parts in the points B, C, C', B'. We support the rod in these four points by struts, the other extremities of which are fastened to the framework. The points B, C, C', B' are fixed, as they are sustained by the struts: hence a weight suspended from P, which is to break the bridge, must be sufficiently strong to break a piece C C′, which is secured at the ends; the rod A A′ would have been broken with 38 lbs., hence 190 lbs. would be necessary to break C C′. There is a similar beam on the other side of the bridge, and therefore to break the bridge 380 lbs. would be necessary, but this force must be applied exactly at the centre of C C′; and if the weights be spread over any considerable length, a heavier load will be necessary. In fact, if I were to distribute the weight uniformly over the distance C C′, it appears from Art. 408 that double the load would be necessary to produce fracture.

430. We shall now break this model. I place 18 stone upon it ranged uniformly, and the cathetometer tells me that the bridge only deflects 0″·1, and that its elasticity is not injured. Placing the tray in position, and loading the bridge by this means, I find with a weight of 2 cwt. that there is a deflection of 0″·15; with 4 cwt. the deflection amounts to 0′·72. We therefore infer that the bridge is beginning to yield, and the clamps give way when the load is increased to 500 lbs.

A BRIDGE WITH TWO TIES.

431. It might happen that circumstances would not make it convenient to obtain points of support below the bridge on which to erect the struts. In such a case, if suitable positions for ties can be obtained, a bridge of the form represented in Fig. 60 may be used.

A D is a horizontal rod of pine 40″ × 0″·5 × 0″·5; it is trisected in the points B and F, from which points the ties B E and C F are secured to the upper parts of the framework. A D is then supported in the points B and C, which may therefore be regarded as fixed points. Hence, for the reasons we have already explained, the strength of the bridge should be increased nearly threefold. Remembering that the bridge has two beams we know it would require about 70 lbs. or 80 lbs. to produce fracture without the ties, and therefore we might expect that over 200 lbs. would be necessary when the beams were supported by the ties. I perform the experiment, and you see the bridge yields when the load reaches 194 lbs.: this is somewhat less than the amount we had calculated; the reason being, I think, that one of the clamps slipped before fracture.

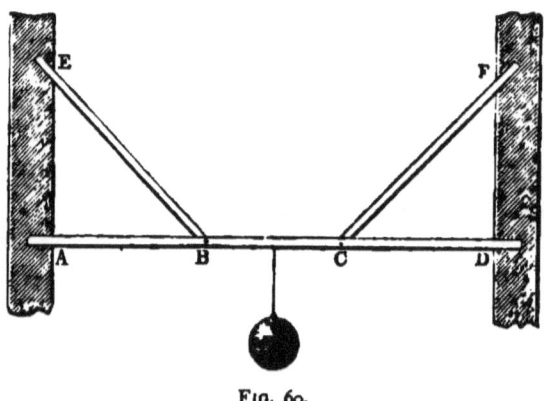

FIG. 60.

A SIMPLE FORM OF TRUSS.

432. It is often not convenient, or even possible, to sustain a bridge by the methods we have been considering. It is desirable therefore to inquire whether we cannot arrange some plan of strengthening a beam, by giving to it what shall be equivalent to an increase of depth.

433. We shall only be able to describe here some very simple methods for doing this. Superb examples are to be found in railway bridges all over the country, but the full investigation of these complex structures is a problem of no little difficulty, and one into which it would be quite

A SIMPLE FORM OF TRUSS.

beyond our province to enter. We shall, however, show how by a judicious combination of several parts a structure can offer sufficient resistance. The most complex lattice girder is little more than a network of ties and struts.

434. Let A B (Fig. 61) be a rod of pine $40'' \times 0''\cdot 5'' \times 0'''\cdot 5$, secured at each end. We shall suppose that the load is applied at the two points G and H, in the manner shown in the figure. The load which a bridge must bear when a train passes over it is distributed over a distance equal to the length of the train, and the weight of the bridge itself is of course arranged along the entire span; hence the load which

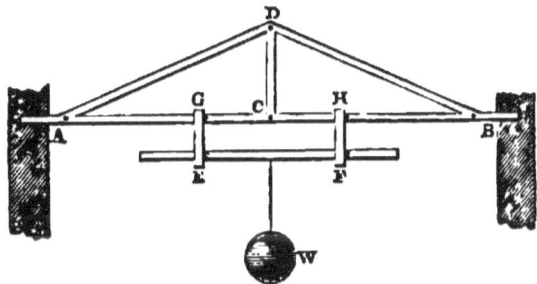

FIG. 61.

a bridge bears is at all times more or less distributed and never entirely concentrated at the centre in the manner we have been considering. In the present experiment we shall apply the breaking load at the two points G and H, as this will be a variation from the mode we have latterly used. E F is an iron bar supported in the loops E G and F H. Let us first try what weight will break the beam. Suspending the tray from E F, I find that a load of 48 lbs. is sufficient; much less would have done had not the ends been clamped. We have already applied a load in this manner in Art. 406.

435. You observed that the beam, as usual, deflected before it broke; if we could prevent deflection we might reasonably expect to increase the strength. Thus if we support the centre of the beam C, deflection would be prevented. This can be done very simply. We clamp the pieces D A, D B, D C, on a similar beam, and it is evident that C cannot descend so long as the joints at A, B, D, C remain firmly secured. We now find that even with a weight of 112 lbs.

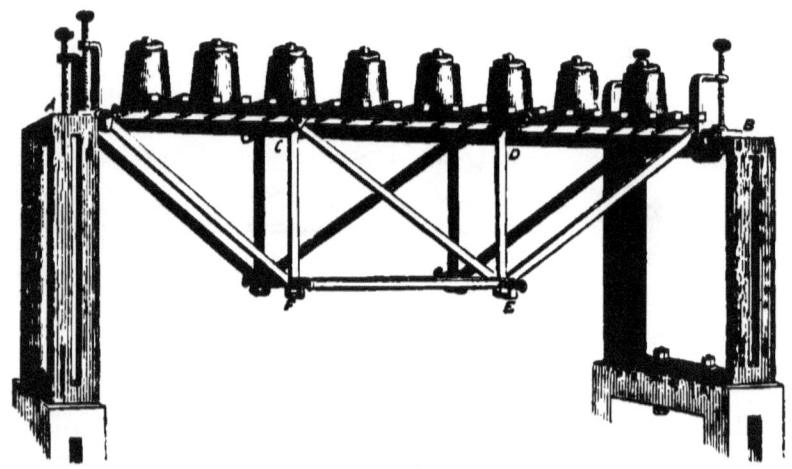

FIG. 62.

in the tray, the bar is unbroken. An arrangement of this kind is frequently employed in engineering, for it seems to be able to bear more than double the load which is sufficient to break the unsupported beam.

436. Two frames of this kind, with a roadway laid between them, would form a bridge, or if the frames were turned upside down they would answer equally well, though of course in this case D A and D B would become ties, and D C a strut, but a better arrangement for a bridge will be next described.

XIII.] THE TRUSSED BRIDGE. 215

THE WYE BRIDGE.

437. An instructive bridge was erected by the late Sir I. Brunel over the Wye, for the purpose of carrying a railway. The essential parts of the bridge are represented in the model shown in Fig. 62, which as before is made of slips of pine clamped together.

438. Our model is composed of two similar frames, one of which we shall describe. A B is a rod of pine $48'' \times 0''{\cdot}5 \times 0''{\cdot}5$, supported at each extremity. This rod is sustained at its points of trisection D, C by the uprights D E and C F, while E and F are supported by the rods B E, F E, and A F; the rectangle D E F C is stiffened by the piece C E, and it would be proper in an actual structure to have a piece connecting D and F, but it has not been introduced into the model.

439. We shall understand the use of the diagonal C E by an inspection of Fig. 63. Suppose the quadrilateral A B C D be formed of four pieces of wood hinged at the corners. It is evident that this quadrilateral can be deformed by pressing A and C together, or by pulling them asunder. Even if there were actual joints at the corners, it would be almost impossible to make the quadrilateral stiff by the strength of the joints.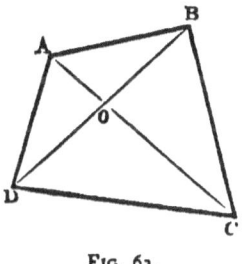

FIG. 63.

You see this by the frame which I hold in my hand; the pieces are clamped together at the corners, but no matter how tightly I compress the clamps, I am able with the slightest exertion to deform the figure.

440. We must therefore look for some method of stiffening the frame. I have here a triangle of three pieces, which

have been simply clamped together at the corners; this triangle is unalterable in form; in fact, since it is impossible to make two different triangles with the same three sides, it is evident the triangle cannot be deformed. This points to a guiding principle in all bridgework. The quadrilateral is not stiff because innumerable different quadrilaterals can be made with the same four sides. But if we draw the diagonal A C of the quadrilateral it is divided into two triangles, and hence when we attach to the quadrilateral, which has been clamped at the four corners, an additional piece in the direction of one of the diagonals, it becomes unalterable in shape.

441. In Fig. 63 we have drawn the two diagonals A C and B D: one would be theoretically sufficient, but it is desirable to have both, and for the following reason. If I pull A and C apart, I stretch the diagonal A C and compress B D. If I compress A and C together, I compress the line A C and extend B D; hence in one of these cases A C is a tie, and in the other it is a strut. It therefore follows that in all cases one of the diagonals is a tie, and the other a strut. If then we have only one diagonal, it is called upon to perform alternately the functions of a tie and of a strut. This is not desirable, because it is evident that a piece which may act perfectly as a tie may be very unsuitable for a strut, and *vice versâ*. But if we insert both diagonals we may make both of them ties, or both of them struts, and the frame must be rigid. Thus for example, I might make A C and B D slender bars of wrought iron, which form admirable ties, though quite incapable of acting as struts.

442. What we have said with reference to the necessity for dividing a quadrilateral figure into triangles applies still more to a polygon with a large number of sides, and we may lay down the general principle that every such piece of framework should be composed of triangles.

XIII.] THE TRUSSED BRIDGE. 217

443. Returning to Fig. 62, we see the reason why the rectangle E D C F should have one or both of its diagonals introduced. A load placed, for example, at D would tend to depress the piece D E, and thus deform the rectangle, but when the diagonals are introduced this deformation is impossible.

444. Hence one of these frames is almost as strong as a beam supported at the points C and D, and therefore, from the principles of Art. 388, its strength is three times as great as that of an unsupported beam.

445. The two frames placed side by side and carrying a roadway form an admirable bridge, quite independent of any external support, except that given by the piers upon which the extremities of the frames rest. It would be proper to connect the frames together by means of braces, which are not, however, shown in the figure. The model is represented as carrying a uniform load in contradistinction to Fig. 58, where the weight is applied at a single point.

446. With eight stone ranged along it, the bridge of Fig. 62 did not indicate an appreciable deflection.

LECTURE XIV.

THE MECHANICS OF A BRIDGE.

Introduction.—The Girder.—The Tubular Bridge.—The Suspension Bridge.

INTRODUCTION.

447. PERHAPS it may be thought that the structures we have been lately considering are not those which are most universally used, and that the bridges which are generally referred to as monuments of engineering skill are of quite a different construction. Every one is familiar with the arch, and most of us have seen suspension bridges and the celebrated Menai tube. We must therefore allude further to some of these structures, and this we propose to do in the present lecture. It will only be possible to take a very slight survey of an extensive subject to which elaborate treatises have been devoted.

We shall first give a brief account of the use of iron in the arts of construction. We shall then explain simply the principle of the tubular bridge, and also of the suspension bridge. The more complex forms are beyond our scope.

THE GIRDER.

448. A horizontal beam supported at each end, and perhaps at intermediate points, and designed to support a heavy load is called a *girder*. Those rods upon which we have performed experiments, the results of which have been given in Table XXIV., are small girders; but the term is generally understood to relate to structures of iron : the greatest girders for railway bridges are made of bars or plates of iron riveted together.

449. We shall first consider the application of *cast* iron to girders, and show what form they should assume.

450. A beam of cast iron, supposing its section to be rectangular, has its strength determined by the same laws as the beams of pine. Thus, supposing the section of two beams to be the same, their strengths are inversely proportional to their lengths, and the strength of a beam placed edgewise is to its strength placed flatwise in the proportion of the greater dimension of its section to the less dimension. These laws determine the strength of every rectangular beam of cast iron when that of one beam is known, and we must perform an experiment in order to find the breaking load in a particular case.

451. I take here a piece of cast iron, which is 2' long, and 0"·5 × 0"·5 in section. I support this beam at each end upon a frame; the distance between the supports is 20". I attach the tray to the centre of the beam and load it with weights. The ends of the beam rest freely upon the supports, but I have taken the precaution of tying each end by a piece of wire, so that they may not fly about when the fracture occurs. Loading the tray, I find that with 280 lbs. the crash comes.

452. Let us compare this result with No. 8 of Table XXIV.

(p. 190). There we find that a piece of pine, the same size as the cast iron, was broken with 36 lbs.: the ratio of 280 to 36 is nearly 8, so that the beam of cast iron is about 8 times as strong as the piece of pine of the same size. This result is a little larger than we would have inferred from an examination of tables of the strength of large bars of cast iron; the reason may be that a very small casting, such as this bar, is stronger in proportion than a larger one, owing to the iron not being so uniform throughout the larger mass.

453. I hold here a bar of cast iron 12″ long and 1″ × 1″ in section. I have not sufficient weights at hand to break it, but we can compute how much would be necessary by our former experiment.

454. In the first place a bar 12″ long, and 0″·5 × 0″·5 of section, would require 20 × 280 ÷ 12 = 467 lbs. by the law that the strength is inversely as the length. We also know that one beam 12″ × 1″ × 1″ is just as strong as two beams 12″ × 1″ × 0″·5, each placed edgewise; each of these latter beams is twice as strong as 12″ × 1″ × 0″·5 placed flatwise, because the strength when placed edgewise is to the strength when placed flatwise, as the depth to the breadth, that is as 2 to 1: hence the original beam is four times as strong as one beam 12″ × 1″ × 0″·5 placed flatwise: but this last beam is twice as strong as a beam 12″ × 0″·5 × 0″·5, and hence we see that a beam 12″ × 1″ × 1″ must be 8 times as strong as a beam of 12″ × 0″·5 × 0″·5, but this last beam would require a load of 467 lbs. to break it, and hence the beam of 12″ × 1″ × 1″ would require 467 × 8 = 3736 lbs. to produce fracture. This amounts to more than a ton and a half.

455. It is a rule sometimes useful to practical men that

a cast iron bar one foot long by one inch square would break with about a ton weight. If the iron be of the same quality as that which we have used, this result is too small, but the error is on the safe side; the real strength will then be generally a little greater than the strength calculated from this rule. What we have said (Art. 403) with reference to the precaution for safety in bars of wood applies also to cast iron. The load which the beam has to bear in ordinary practice should only be a small fraction of that which would break it.

456. In making any description of girder it is desirable on very special grounds that as little material as possible be uselessly employed. It will of course be remembered that a girder has to support its own weight, besides whatever may be placed upon it: and if the girder be massive, its own weight is a serious item. Of two girders, each capable of bearing the same *total* load, the lighter, besides employing less material, will be able to bear a greater weight placed upon it. It is therefore for a double reason desirable to diminish the weight. This remark applies especially to such a material as cast iron, which can be at once given the form in which it shall be capable of offering the greatest resistance.

457.--The principles which will guide us in ascertaining the proper form to give a cast iron girder, are easily deduced from what we have laid down in Lectures XI. and XII. We have seen that depth is very desirable for a strong beam. If therefore we strive to attain great depth in a light beam, the beam must be very thin. Now an extremely thin beam will not be safe. In the first place it would be flexible and liable to displacement sideways; and, in the second place, there is a still more fatal difficulty. We have shown that when a beam of wood is supporting a weight, the fibres at the bottom of the beam are extended, the tendency being to tear them (Art.376). The fibres on the top of the beam are compressed, while the centre of the beam is in its natural state. The condition of strain in a cast-iron beam is precisely similar; the bottom portions are in a state of extension, while the top is compressed. If therefore a beam be very thin, the material at the lower part may not be sufficient to withstand the forces of extension, and fracture is produced. To obviate this, we strengthen the bottom of the beam by placing extra material there. Thus we are led to the idea of a thin beam with an excess of iron at the bottom.

FIG. 64.

458. E F (Fig. 64) is the thin iron beam along the bottom of which is the stout flange shown at C D; rupture cannot commence at the bottom unless this flange be torn asunder; for until this happens it is clear that fracture cannot begin to attack the upper and slender part of the beam E F.

459. But the beam is in a state of compression along its upper side, just as in the wooden beams which we have already considered. If therefore the upper parts were not powerful enough to resist this compression, they would be crushed, and the beam would give way. The remedy for this source of weakness is obvious; a second flange runs along the top of the beam, as shown at A B. If this be strong enough to resist the compression, the stability of the beam is ensured.

460. The upper flange is made very much smaller than

the lower one, in consequence of a property of cast iron. This metal is more capable of resisting forces of compression than forces of extension, and it is only necessary to use one-sixth of the iron on the upper flange that is required for the lower. When the section has been thus proportioned, the beam is equally strong at both top and bottom; adding material to either flange without strengthening the other, will not benefit the girder, but will rather prove a source of weakness, by increasing the weight which has to be supported.

461. I have here a small girder made of what we are familiar with under the name of "tin," but which is of course sheet iron thinly covered over with tin. It has the shape shown in Fig. 64, and it is 12" long. I support it at each end, and you see it bears two hundred weight without apparent deflection.

THE TUBULAR BRIDGE.

462. I shall commence the description of the principle of this bridge by performing some experiments upon a tube, which I hold in my hand. The tube is square, 1" × 1" in section, and 38" long. It is made of "tin," and weighs rather less than a pound.

463. Here is a solid rod of iron of the same length as the tube, but containing considerably more metal. This is easily verified by weighing the tube and the rod one against the other. I shall regard them as two girders, and experiment upon their strength, and we shall find that, though the tube contains less substance than the rod, it is much the stronger.

464. I place the rod on a pair of supports about 3' apart; I then attach the tray to the middle of the rod: 14 lbs. produce a deflection of 0″·51, and 42 lbs. bends down the rod through 3″·18. This is a large deflection; and when

I remove the load, the rod only returns through 1″·78, thus showing that a permanent deflection of 1″·40 is produced. This proves that the rod is greatly injured, and demonstrates its unsuitability for a girder.

465. Next we place the tube upon the same supports, and treat it in the same manner. A load of 56 lbs. only produces a deflection of 0″·09, and, when this load is removed, the tube returns to its original position: this is shown by the cathetometer, for a cross is marked on the tube, and I bring the image of it on the horizontal wire of the telescope before the load of 56 lbs. is placed in the tray. When the load is removed, I see that the cross returns exactly to where it was before, thus proving that the elasticity of the tube is unimpaired. We double the load, thus placing 1 cwt. in the tray, the deflection only reaches 0″·26, and, when the load is removed, the tube is found to be permanently deflected by a quantity, at all events not greater than 0″·004; hence we learn that the tube bears easily and without injury a load more than twice as great as that which practically destroyed a rod of wrought iron, containing more iron than the tube. We load the tube still further by placing additional weights in the tray, and with 140 lbs. the tube breaks; the fracture has occurred at a joint which was soldered, and the real breaking strength of the tube, had it been continuous, is doubtless far greater. Enough, however, has been borne to show the increase of strength obtained by the tubular form.

466. We can explain the reason of this remarkable result by means of Fig. 64. Were the thin portion of the girder E F made of two parts placed side by side, the strength would not be altered. If we then imagine the flange A B widened to the width of C D, and the two parts which form E F opened out so as to form a tube,

the strength of the girder is still retained in its modified form.

467. A tube of rectangular section has the advantage of greater depth than a solid rod of the same weight; and if the bottom of the tube be strong enough to resist the extension, and the top strong enough to resist the compression, the girder will be stiff and strong.

468. In the Menai Tubular Bridge, where a gigantic tube supported at each end bridges over a span of four hundred and sixty feet, special arrangements have been made for strengthening the top. It is formed of cells, as wrought iron disposed in this way is especially adapted for resisting compression.

469. We have only spoken of rectangular tubes, but it is equally true for tubes of circular or other sections that when suitably constructed they are stronger than the same quantity of material, if made into a solid rod.

470. We find this principle in nature; bones and quills are often found to be hollow in order to combine lightness with strength, and the stalks of wheat and other plants are tubular for the same reason.

THE SUSPENSION BRIDGE.

471. Where a great span is required, the suspension bridge possesses many advantages. It is lighter than a girder bridge of the same span, and consequently cheaper, while its singular elegance contrasts very favourably with the appearance of more solid structures. On the other hand, a suspension bridge is not so well suited for railway traffic as the lattice girder.

472. The mechanical character of the suspension bridge is simple. If a rope or a chain be suspended from two points to which its ends are attached, the chain hangs in a

certain curve known to mathematicians as the *catenary*. The form of the *catenary* varies with the length of the rope, but it would not be possible to make the chain lie in a straight line between the two points of support, for reasons pointed out in Art. 20. No matter how great be the force applied, it will still be concave. When the chain is stretched until the depression in the middle is small compared with the distance between the points of support, the curve though always a catenary, has a very close resemblance to the *parabola*.

473. In Fig. 65 a model of a suspension bridge is shown. The two chains are fixed one on each side at the points E and F; they then pass over the piers A, D, and bridge a span of nine feet. The vertical line at the centre B C shows the greatest amount by which the chain has deflected from the horizontal A D. When the deflection of the middle of the chain is about one-tenth part of A D, the curve A C D becomes for all practical purposes a parabola. The roadway is suspended by slender iron rods from the chains, the lengths of the suspension rods being so regulated as to make it nearly horizontal.

474. The roadway in the model is laden with 8 stone weights. We have distributed them in this manner in order to represent the permanent load which a great suspension bridge has to carry. The series of weights thus arranged produces substantially the same effect as if it were actually distributed uniformly along the length. In a real suspension bridge the weight of the chain itself adds greatly to the tension.

475. We assume that the chain hangs in the form of a parabola, and that the load is uniformly ranged along the bridge. The tension upon the chains is greatest at their highest points, and least at their lowest points, though

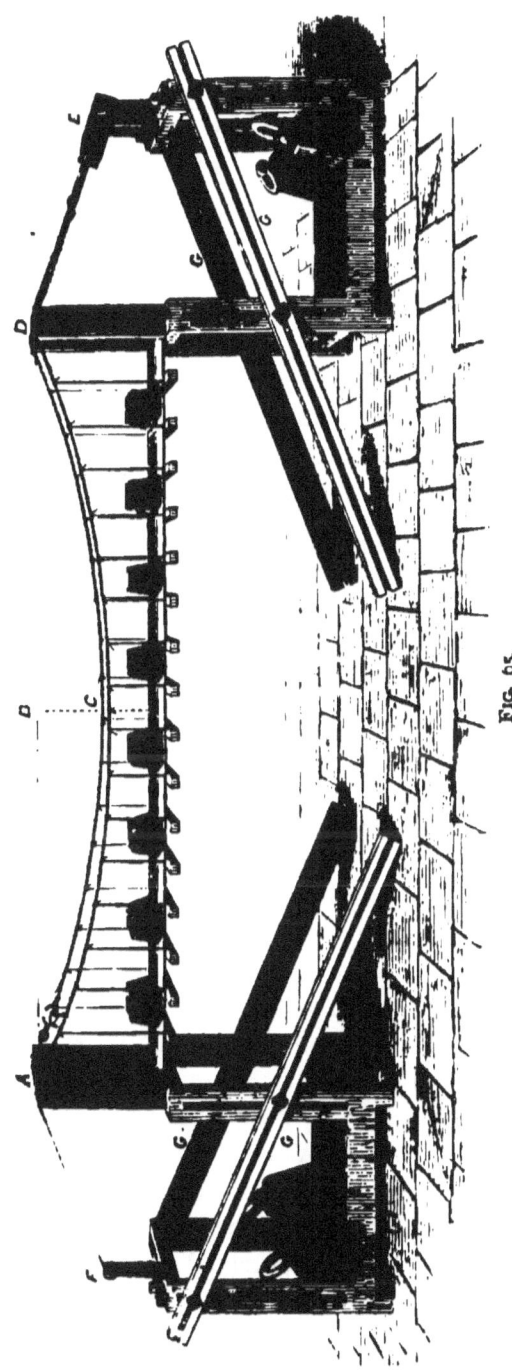

Fig. 65.

the difference is small. The amount of the tension can be calculated when the load, span, and deflection are known. We cannot give the steps of the calculation, but we shall enunciate the result.

476. The magnitude of the tension at the lowest point C of each chain is found by multiplying the total weight (including chains, suspension rods, and roadway) by the span, and dividing the product by sixteen times the deflection.

The tension of the chain at the highest point A exceeds that at the lowest point C, by a weight found by multiplying the total load by the deflection, and dividing the product by twice the span.

477. The total weight of roadway, chains, and load in the model is 120 lbs.; the deflection is 10″, the span 108″; the product of the weight and span is 12,960; sixteen times the deflection is 160; and, therefore, the tension at the point C is found, by dividing 12,960 by 160, to be 81 lbs.

To find the tension at the point A, we multiply 120 by 10, and divide the product by 216; the quotient is nearly 6. This added to 81 lbs. gives 87 lbs. for the tension on the chain at A.

478. One chain of the model is attached to a spring-balance at A; by reference to the scale we see the tension indicated to be 90 lbs.: a sufficiently close approximation to the calculated tension of 87 lbs.

479. A large suspension bridge has its chains strained by an enormous force. It is therefore necessary that the ends of these chains be very firmly secured. A good attachment is obtained by anchoring the chain to a large iron anchor imbedded in solid rock.

480. In Art. 45 we have pointed out how the dimensions of the tie rod could be determined when the tension was

known. Similar considerations will enable us to calculate the size of the chain necessary for a suspension bridge when we have ascertained the tension to which it will be subjected.

481. We can easily determine by trial what effect is produced on the tension of the chain, by placing a weight upon the bridge in addition to the permanent load. Thus an additional stone weight in the centre raises the tension of the spring-balance to 100 lbs.; of course the tension in the other chain is the same: and thus we find a weight of 14 lbs. has produced additional tensions of 10 lbs. each in the two chains. With a weight of 28 lbs. at the centre we find a strain of 110 lbs. on the chain.

482. These additional weights may be regarded as analogous to the weights of the vehicles which the suspension bridge is required to carry. In a large suspension bridge the tension produced by the passing loads is only a small fraction of the permanent load.

LECTURE XV.

THE MOTION OF A FALLING BODY.

Introduction.—The First Law of Motion.—The Experiment of Galileo from the Tower of Pisa.—The Space is proportional to the Square of the Time.—A Body falls 16' in the First Second.—The Action of Gravity is independent of the Motion of the Body.—How the Force of Gravity is defined.—The Path of a Projectile is a Parabola.

INTRODUCTION.

483. Kinetics is that branch of mechanics which treats of the action of forces in the production of motion. We shall find it rather more difficult than the subjects with which we have been hitherto occupied; the difficulties in kinetics arise from the introduction of the element of *time*, into our calculations. The principles of kinetics were unknown to the ancients. Galileo discovered some of its truths in the seventeenth century; and, since his time, the science has grown rapidly. The motion of a falling body was first correctly apprehended by Galileo; and with this subject we can appropriately commence.

THE FIRST LAW OF MOTION.

484. Velocity, in ordinary language, is supposed to convey a notion of rapid motion. Such is not precisely the

LECT. XV.] THE FIRST LAW OF MOTION. 231

meaning of the word in mechanics. By velocity is merely meant the *rate* at which a body moves, whether the rate be fast or be slow. This rate is most conveniently measured by the number of feet moved over in one second. Hence when it is said the velocity of a body is 25, it is meant that if the body continued to move for one second with its velocity unaltered, it would in that time have moved over 25 feet.

485. The first law of motion may be stated thus. *If no force act upon a body, it will, if at rest, remain for ever at rest; or if in motion, it will continue for ever to move with a uniform velocity.* We know this law to be true, and yet no one has ever seen it to be true for the simple reason that we cannot realise the condition which it requires. We cannot place a body in the condition of being unacted upon by any forces. But we may convince ourselves of the truth of the law by some such reasoning as the following. If a stone be thrown along the road, it soon comes to rest. The body leaves the hand with a certain initial velocity and is not further acted upon by it. Hence, if no other force acted on the stone, we should expect, if the first law be true, that it would continue to run on for ever with the original velocity at the moment of leaving the hand. But other forces do act upon the stone; the attraction of the earth pulls it down; and, when it begins to bound and roll upon the ground, friction comes into operation, deprives the stone of its velocity, and brings it to rest. But let the stone be thrown upon a surface of smooth ice; when it begins to slide, the force of gravity is counteracted by the reaction of the ice: there is no other force acting upon the stone except friction, which is small. Hence we find that the stone will run on for a considerable distance. It requires but little effort of the imagination to suppose a lake whose

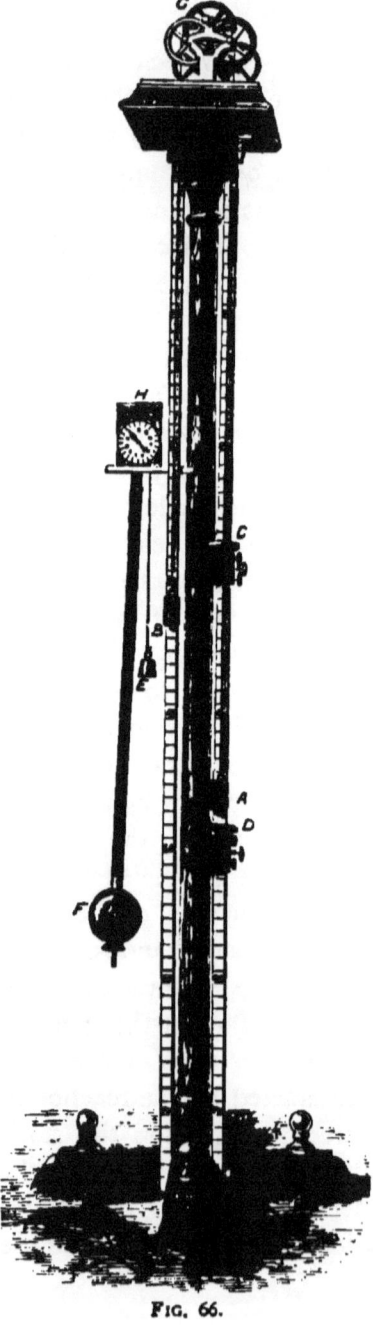

Fig. 66.

[LECT.

surface is an infinite plane, perfectly smooth, and that the stone is perfectly smooth also. In such a case as this the first law of motion amounts to the assertion that the stone would never stop.

486. We may, in the lecture room, see the truth of this law verified to a certain extent by Atwood's machine (Fig. 66). This machine has been devised for the purpose of investigating the laws of motion by actual experiment. It consists principally of a pulley C, mounted so that its axle rests upon two pairs of wheels, as shown in the figure; it being the object of this contrivance to enable the wheel to revolve with the utmost freedom. A pair of equal weights A, B, are attached by a silken thread, which passes over the pulley; each of the weights is counterbalanced by the other: so that when the two are in motion, we may consider either as a body

xv.] THE EXPERIMENT OF GALILEO. 233

not acted upon by any forces, and it will be found that it moves uniformly, as far as the size of the apparatus will permit.

487. If we try to conceive a body free in space, and not acted upon by any force, it is more natural to suppose that such a body, when once started, should go on moving uniformly for ever, than that its velocity should be altered. The true proof of the first law of motion is, that all consequences properly deduced from it, in combination with other principles, are found to be verified. Astronomy presents us with the best examples. The calculation of the time of an eclipse is based upon laws which in themselves assume the first law of motion; hence, when we invariably find that an eclipse occurs precisely at the moment for which it has been predicted, we have a splendid proof of the sublime truth which the first law of motion expresses.

THE EXPERIMENT OF GALILEO FROM THE TOWER OF PISA.

488. The contrast between heavy bodies and light bodies is so marked that without trial we hardly believe that a heavy body and a light body will fall from the same height in the same time. That they do so Galileo proved by dropping a heavy ball and a light ball together from the top of the Leaning Tower at Pisa. They were found to reach the ground simultaneously. We shall repeat this experiment on a scale sufficiently reduced to correspond with the dimensions of the lecture room.

489. The apparatus used is shown in Fig. 67. It consists of a stout framework supporting a pulley H at a height of about 20 feet above the ground. This pulley carries a rope; one end of the rope is attached to a

Fig. 67.

triangular piece of wood, to which two electro-magnets G are fastened. The electro-magnet is a piece of iron in the form of a horse-shoe, around which is coiled a long wire. The horse-shoe becomes a magnet immediately an electric current passes through the wire; it remains a magnet as long as the current passes, and returns to its original condition the moment the current ceases. Hence, if I have the means of controlling the current, I have complete control of the magnet; you see this ball of iron remains attached to the magnet as long as the current passes, but drops the instant I break the current. The same electric circuit includes both the magnets; each of them will hold up an iron ball F when the current passes, but the moment the current is broken both balls will be released. Electricity travels along a wire with prodigious velocity. It would pass over many thousands of miles in a second; hence the time that it takes to pass through the wires we are employing is quite inappreciable. A piece of thin paper interposed between the magnets and the balls will ensure that they are dropped simultaneously; when this precaution is not taken one or both balls may hesitate a little before commencing to descend. A long pair of wires E, B, must be attached to the magnets, the other ends of the wires communicating with the battery D; the triangle and its load is hoisted up by means of the rope and pulley and the magnets thus carry the balls to a height of 20 feet: the balls we are using weigh about 0·25 lb. and 1 lb. respectively.

490. We are now ready to perform the experiment. I break the circuit; the two balls are disengaged simultaneously; they fall side by side the whole way, and reach the ground together, where it is well to place a cushion to receive them. Thus you see the heavy ball and the

light one each require the same amount of time to fall from the same height.

491. But these balls are both of iron; let us compare together balls made of different substances, iron and wood for example. A flat-headed nail is driven into a wooden ball of about $2''\cdot5$ in diameter, and by means of the iron in the nail I can suspend this ball from one of the magnets; while either of the iron balls we have already used hangs from the other. I repeat the experiment in the same manner, and you see they also fall together. Finally, when an iron ball and a cork ball are dropped, the latter is within two or three inches of its weighty companion when the cushion is reached: this small difference is due to the greater effect of the resistance of the air on the lighter of the two bodies. There can be no doubt that in a vacuum all bodies of whatever size or material would fall in precisely the same time.

492. How is the fact that all bodies fall in the same time to be explained? Let us first consider two iron balls. Take two equal particles of iron: it is evident that these fall in the same time; they would do so if they were very close together, even if they were touching, but then they might as well be in one piece: and thus we should find that a body consisting of two or more iron particles takes the same time to fall as one (omitting of course the resistance of the air). Thus it appears most reasonable that two balls of iron, even though unequal in size, should fall in the same time.

493. The case of the wooden ball and the iron ball will require more consideration before we realise thoroughly how much Galileo's experiment proves. We must first explain the meaning of the word *mass* in mechanics.

494. It is not correct to define *mass* by the introduction

XV.] THE EXPERIMENT OF GALILEO. 237

of the idea of weight, because the *mass* of a body is something independent of the existence of the earth, whereas weight is produced by the attraction of the earth. It is true that weight is a convenient means of measuring *mass*, but this is only a consequence of the property of gravity which the experiment proves, namely, that the attraction of gravity for a body is proportional to its *mass*.

495. Let us select as the unit of mass the mass of a piece of platinum which weighs 1 lb. at London; it is then evident that the mass of any other piece of platinum should be expressed by the number of pounds it contains: but how are we to determine the mass of some other substance, such as iron? A piece of iron is defined to have the same mass as a piece of platinum, if the same force acting on either of the bodies for the same time produces the same velocity. This is the proper test of the equality of masses. The mass of any other piece of iron will be represented by the number of times it contains a piece equal to that which we have just compared with the platinum; similarly of course for other substances.

496. The magnitude of a force acting for the time unit is measured by the product of the mass set in motion and the velocity which it has acquired. This is a truth established, like the first law of motion, by indirect evidence.

497. Let us apply these principles to explain the experiment which demonstrated that a ball of wood and a ball of iron fall in the same time. Forces act upon the two bodies for the same time, but the magnitudes of the forces are proportional to the mass of each body multiplied into its velocity, and, since the bodies fall simultaneously, their velocities are equal. The forces acting upon the bodies are therefore proportional to their masses; but the force acting on each body is the attraction of the earth, therefore, the

gravitation to the earth of different bodies is proportional to their masses.

498. We may here note the contrast between the attraction of gravitation and that of a magnet. A magnet attracts iron powerfully and wood not at all; but the earth draws all bodies with forces depending on their masses and their distances, and not on their chemical composition.

THE SPACE DESCRIBED BY A FALLING BODY IS PROPORTIONAL TO THE SQUARE OF THE TIME.

499. We have next to discover the law by which we ascertain the distance a body falling from rest will move in a given time; it is not possible to experiment directly upon this subject, as in two seconds a body will drop 64 feet and acquire an inconveniently large velocity; we can, however, resort to Atwood's machine (Fig. 66) as a means of diminishing the motion. For this purpose we require a clock with a seconds pendulum.

500. On one of the equal cylinders A I place a slight brass rod, whose weight gives a preponderance to A, which will consequently descend. I hold the loaded weight in my hand, and release it simultaneously with the tick of the pendulum. I observe that it descends 5" before the next tick. Returning the weight to the place from whence it started, I release it again, and I find that at the second tick of the pendulum it has travelled 20". Similarly we find that in three seconds it descends 45". It greatly facilitates these experiments to use a little stage which is capable of being slipped up and down the scale, and which can be clamped to the scale in any position. By actually placing the stage at the distance of 5", 20", or 45" below the point from which the weight starts, the coincidence of the tick of the pendulum

xv.] FALL OF A BODY IN THE FIRST SECOND. 239

with the tap of the weight on its arrival at the stage is very marked.

501. These three distances are in the proportion of 1, 4, 9; that is, as the squares of the numbers of seconds 1, 2, 3. Hence we may infer that *the distance traversed by a body falling from rest is proportional to the square of the time.*

502. The motion of the bodies in Atwood's machine is much slower than the motion of a body falling freely, but the law just stated is equally true in both cases so that in a free fall the distance traversed is proportional to the square of the time. Atwood's machine cannot directly tell us the distance through which a body falls in one second. If we can find this by other means, we shall easily be able to calculate the distance through which a body will fall in any number of seconds.

A BODY FALLS 16' IN THE FIRST SECOND.

503. The apparatus by which this important truth may be demonstrated is shown in Fig. 67. A part of it has been already employed in performing the experiment of Galileo, but two other parts must now be used which will be briefly explained.

504. At A a pendulum is shown which vibrates once every second; it need not be connected with any clockwork to sustain the motion, for when once set vibrating it will continue to swing some hundreds of times. When this pendulum is at the middle of its swing, the bob just touches a slender spring, and presses it slightly downwards. The electric current which circulates about the magnets G (Art. 489) passes through this spring when in its natural position; but when the spring is pressed down by the pendulum, the current is interrupted. The consequence is that, as the

pendulum swings backwards and forwards, the current is broken once every second. There is also in the circuit an electric alarm bell c, which is so arranged that, when the current passes, the hammer is drawn from the bell; but, when the current ceases, a spring forces the hammer to strike the bell. When the circuit is closed, the hammer is again drawn back. The pendulum and the bell are in the same circuit, and thus every vibration of the pendulum produces a stroke of the bell. We may regard the strokes from the bell as the ticks of the pendulum rendered audible to the whole room.

505. You will now understand the mode of experimenting. I draw the pendulum aside so that the current passes uninterruptedly. An iron ball is attached to one of the electro-magnets, and it is then gently hoisted up until the height of the ball from the ground is about 16'. A cushion is placed on the floor in order to receive the falling body. You are to look steadily at the cushion while you listen for the bell. All being ready, the pendulum, which has been held at a slight inclination, is released. The moment the pendulum reaches the middle of its swing it touches the spring, rings the bell, breaks the current which circulated around the magnet, and as there is now nothing to sustain the ball, it drops down to the cushion; but just as it arrives there, the pendulum has a second time broken the electric circuit, and you observe the falling of the ball upon the cushion to be identical with the second stroke of the bell. As these strokes are repeated at intervals of a second, it follows that the ball has fallen 16' in one second. If the magnet be raised a few feet higher, the ball may be seen to reach the cushion after the bell is heard. If the magnet be lowered a few feet, the ball reaches the cushion before the bell is heard.

506. We have previously shown that the space is proportional to the square of the time. We now see that when the time is one second, the space is 16 feet. Hence if the time were two seconds, the space would be $4 \times 16 = 64$ feet; and in general the space in feet is equal to 16 multiplied by the square of the time in seconds.

507. By the help of this rule we are sometimes enabled to ascertain the height of a perpendicular cliff, or the depth of a well. For this purpose it is convenient to use a stopwatch, which will enable us to measure a short interval of time accurately. But an ordinary watch will do nearly as well, for with a little practice it is easy to count the beats, which are usually at the rate of five a second. By observing the number of beats from the moment the stone is released till we see or hear its arrival at the bottom, we determine the time occupied in the act of falling. The square of the number of seconds (taking account of fractional parts) multiplied by 16 gives the depth of the well or the height of the cliff in feet, provided it be not high.

THE ACTION OF GRAVITY IS INDEPENDENT OF THE MOTION OF THE BODY.

508. We have already learned that the effect of gravity does not depend upon the actual chemical composition of the body. We have now to learn that its effect is uninfluenced by any motion which the body may possess. Gravity pulls a body down 16′ per second, if the body starts from rest. But suppose a stone be thrown upwards with a velocity of 20 feet, where will it be at the end of a second? Did gravity not act upon the stone, it would be at a height of 20 feet. The principle we have stated tells us that gravity will draw this stone towards the earth through a distance of

R

16′, just as it would have done if the stone had started from rest. Since the stone ascends 20′ in consequence of its own velocity, and is pulled back 16′ by gravity, it will, at the end of a second, be found at the height of 4′. If, instead of being shot up vertically, the body had been projected in any other direction, the result would have been the same; gravity would have brought the body at the end of one second 16′ nearer the earth than it would have been had gravity not acted. For example, if a body had been shot vertically downwards with a velocity of 20′, it would in one second have moved through a space of 36′.

509. We shall illustrate this remarkable property by an experiment. The principle of doing so is as follows:—Suppose we take two bodies, A and B. If these be held at the same height, and released together, of course they reach the ground at the same instant; but if A, instead of being merely dropped, be projected with a horizontal velocity at the same moment that B is released, it is still found that A and B strike the floor simultaneously.

510. You may very simply try this without special apparatus. In your left hand hold a marble, and drop it at the same instant that your right hand throws another marble horizontally. It will be seen that the two marbles reach the ground together.

511. A more accurate mode of making the experiment is shown by the contrivance of Fig. 68.

In this we have an arrangement by which we ensure that one ball shall be released just as the other is projected. At A B is shown a piece of wood about 2″ thick; the circular portion (2′ radius) on which the ball rests is grooved, so that the ball only touches the two edges and not the bottom of the groove. Each edge of the groove is covered with tinfoil C, but the pieces of tinfoil on the two sides

xv.] THE ACTION OF GRAVITY. 243

must not communicate. One edge is connected with one pole of the battery K, and the other edge with the other pole, but the current is unable to pass until a communication by a conductor is opened between the two edges. The ball G supplies the bridge; it is covered with tinfoil, and therefore, as long as it rests upon the edges, the circuit is

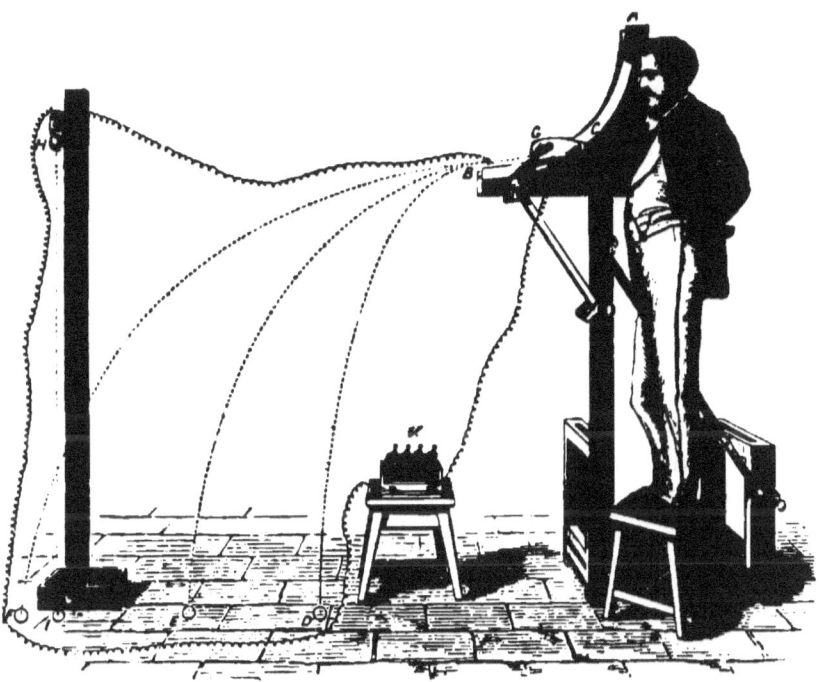

FIG. 68.

complete; the groove is so placed that the tangent to it at the lowest point B is horizontal, and therefore, when the ball rolls down the curve, it is projected from the bottom in a horizontal direction. An india-rubber spring is used to propel the ball; and by drawing it back when embraced by the spring, I can communicate to the missile a velocity which can be varied at pleasure. At H we have an electro-magnet, the wire around which forms part of the circuit we have been considering. This magnet is so placed that a ball suspended from it is precisely at the same height above the floor as the tinned ball is at the moment when it leaves the groove.

512. We now understand the mode of experimenting. So long as the tinned ball G remains on the curve the bridge is complete, the current passes, and the electro-magnet will sustain H, but the moment G leaves the curve, H is allowed to fall. We invariably find that whatever be the velocity with which G is projected, it reaches the ground at the same instant as H arrives there. Various dotted lines in the figure show the different paths which G may traverse; but whether it fall at D, at E, or at F, the time of descent is the same as that taken by H. Of course, if G were not projected horizontally, we should not have arrived at this result: all we assert is that whatever be the motion of a body, it will (when possible) be at the end of a second, sixteen feet nearer the earth than it would have been if gravity had not acted. If the body be projected horizontally, its descent is due to gravity alone, and is neither accelerated nor retarded by the horizontal velocity. What this experiment proves is, that the mere fact of a body having velocity does not affect the action of gravity thereon.

513. Though we have only shown that a horizontal velocity does not affect the action of gravity, yet neither does a velocity in any direction. This is verified, like the first law of motion, by the accordance between the consequences deduced from it and the facts of observation.

514. We may summarize these results by saying that no matter what be the material of which a particle is composed, whether it be heavy or light, moving or at rest, if no force

but gravity act upon the particle for t seconds, it will then be $16t^2$ feet nearer the earth than it would have been had gravity not acted.

515. A proposition which is of some importance may be introduced here. Let us suppose a certain velocity and a certain force. Let the velocity be such that a point starting from A, Fig. 69, would in one second move uniformly to B.

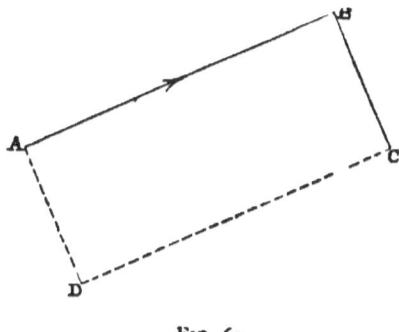

FIG. 69.

Let the force be such that if it acted on a particle originally at rest at A, it would in one second draw the particle to D; if then the force act on a particle having this velocity where will it be at the end of the second? Complete the parallelogram A B C D, and the particle will be found at C. By what we have stated the force will equally discharge its duty whatever be the initial velocity. The force will therefore make the particle move to a distance equal and parallel to A D from whatever position the particle would have assumed, had the force not acted; but had the force not acted, the particle would have been found at B: hence, when the force does act, the particle must be found at C, since B C is equal and parallel to A D.

HOW THE FORCE OF GRAVITY IS DEFINED.

516. From the formula

$$\text{Distance} = 16t^2,$$

we learn that a body falls through 64' in 2 seconds; and as we know that it falls 16' in the first second, it must fall 48' in the next second. Let us examine this. After falling for one second, the body acquires a certain velocity, and with that velocity it commences the next second. Now, according to what we have just seen, gravity will act during the next second quite independently of whatever velocity the body may have previously had. Hence in the second second gravity pulls the body down 16', but the body moves altogether through 48'; therefore it must move through 32' in consequence of the velocity which has been impressed upon it by gravity during the first second. We learn by this that when gravity acts for a second, it produces a velocity such that, if the body be conceived to move uniformly with the velocity acquired, the body would in one second move over 32'.

517. In three seconds the body falls 144', therefore in the third second it must have fallen

$$144' - 64' = 80';$$

but of this 80' only 16' could be due to the action of gravity impressed during that second; the rest,

$$80' - 16' = 64',$$

is due to the velocity with which the body commenced the third second.

518. We see therefore that after the lapse of two seconds gravity has communicated to the body a velocity of 64' per second; we should similarly find, that at the end of the third second, the body has a velocity of 96', and in general at the

xv.] THE PATH OF A PROJECTILE. 247

end of t seconds a velocity of $32t$. Thus we illustrate the remarkable law that *the velocity developed by gravity is proportional to the time.*

519. This law points out that the most suitable way of measuring gravity is by the velocity acquired by a falling body at the end of one second. Hence we are accustomed to say that g (as gravity is generally designated) is 32. We shall afterwards show in the lecture on the pendulum (XVIII.) how the value of g can be obtained accurately. From the two equations, $v = 32t$ and $s = 16t^2$ it is easy to infer another very well known formula, namely, $v^2 = 64s$.

THE PATH OF A PROJECTILE IS A PARABOLA.

520. We have already seen, in the experiments of Fig. 68, that a body projected horizontally describes a curved path on its way to the ground, and we have to determine the geometrical nature of the curve. As the movement is rapid, it is impossible to follow the projectile with the eye so as to ascertain the shape of its path with accuracy; we must therefore adopt a special contrivance, such as that represented in Fig. 70.

B C is a quadrant of wood 2" thick; it contains a groove, along which the ball B will run when released. A series of cardboard hoops are properly placed on a black board, and the ball, when it leaves the quadrant, will pass through all these hoops without touching any, and finally fall into a basket placed to receive it. The quadrant must be secured firmly, and the ball must always start from precisely the same place. The hoops are easily adjusted by trial. Letting the ball run down the quadrant two or three times, we can see how to place the first hoop in its right position, and secure it by drawing pins; then by a few more trials

248 EXPERIMENTAL MECHANICS. [LECT.

the next hoop is to be adjusted, and so on for the whole eight.

521. The curved line from the bottom of the quadrant, which passes through the centres of the hoops, is the path in

Fig. 70.

which the ball moves; this curve is a parabola, of which F is the focus and the line A A the directrix.

It is a property of the parabola that the distance of any point on the curve from the focus is equal to its perpendicular distance from the directrix. This is shown in the figure. For example, the dotted line F D, drawn from F to

xv.] THE PATH OF A PROJECTILE. 249

the centre of the lowest hoop D, is equal in length to the perpendicular D P let fall from D on the directrix A A.

522. The direction in which the ball is projected is in this case horizontal, but, whatever be the direction of projection, the path is a parabola. This can be proved mathematically as a deduction from the theorem of Art. 515.

LECTURE XVI.

INERTIA.

Inertia.—The Hammer.—The Storing of Energy.—The Fly-wheel.—The Punching Machine.

INERTIA.

523. A BODY unacted upon by force will continue for ever at rest, or for ever moving uniformly in a straight line. This is asserted by the first law of motion (Art. 485). It is usual to say that *Inertia* is a property of all matter, by which it is meant that matter cannot of itself change its state of rest or of motion. Force is accordingly required for this purpose. In the present chapter we shall discuss some important mechanical considerations connected with the application of force in changing the state of a body from rest or in altering its velocity when in motion. In the next chapter we shall study the application of force in compelling a body to swerve from its motion in a straight line.

524. We have in earlier lectures been concerned with the application of force either to raise a weight or to overcome friction. We have now to consider the application of force

to a body, not for the purpose of raising it, nor for pushing it along against a frictional resistance, but merely for the purpose of generating a velocity. Unfortunately there is a practical difficulty in the way of making the experiments precisely in the manner we should wish. We want to get a mass isolated both from gravitation and from friction, but this is just what we cannot do—that is, we cannot do it perfectly. We have, however, a simple appliance which will be sufficiently isolated for our present purpose. Here is a heavy weight of iron, about 25 lbs., suspended by an iron wire from the ceiling about 32 feet above the floor (see Fig. 82). This weight may be moved to and fro by the hand. It is quite free from friction, for we need not at present remember the small resistance which the air offers. We may also regard the gravity of the weight as neutralized by the sustaining force of the wire, and accordingly as the body now hangs at rest it may for our purposes be regarded as a body unacted upon by any force.

525. To give this ball a horizontal velocity I feel that I must apply force to it. This will be manifest to you all when I apply the force through the medium of an india-rubber spring. If I pull the spring sharply you notice how much it stretches; you see therefore that the body will not move quickly unless a considerable force is applied to it. It thus follows that merely to generate motion in this mass force has been required.

526. So, too, when the body is in motion as it now is I cannot stop it without the exertion of force. See how the spring is stretched and how strong a pull has thus been exerted to deprive the body of motion. Notice also that while a small force applied sufficiently long will always restore the body to rest, yet that to produce rest quickly a large force will be required.

527. It is an universal law of nature that action and reaction are equal and opposite. Hence when any agent acts to set a body in motion, or to modify its motion in any way, the body reacts on the agent, and this force has been called the *Kinetic reaction*.

528. For example. When a railway train starts, the locomotive applies force to the carriages, and the speed generated during one second is added to that produced during the next, and the pace improves. The kinetic reaction of the train retards the engine from attaining the speed it would acquire if free from the train.

THE HAMMER.

529. The hammer and other tools which give a blow depend for their action upon inertia. A gigantic hammer might force in a nail by the mere weight of the head resting on the nail, but with the help of inertia we drive the nail by blows from a small hammer. We have here inertia aiding in the production of a mechanical power to overcome the considerable resistance which the wood opposes to the entrance of the nail. To drive in the nail usually requires a direct force of some hundreds of pounds, and this we are able conveniently to produce by suddenly checking the velocity of a small moving body.

530. The theory of the hammer is illustrated by the apparatus in Fig. 71. It is a tripod, at the top of which, about 9' from the ground, is a stout pulley C; the rope is about 15' long, and to each end of it A and B are weights attached. These weights are at first each 14 lbs. I raise A up to the pulley, leaving B upon the ground; I then let go the rope, and down falls A: it first pulls the slack rope through, and then, when A is about 3' from the ground, the rope becomes tight, B gets a violent chuck and is lifted into

XVI.] THE HAMMER. 253

the air. What has raised B? It cannot be the mere weight of A, because that being equal to B, could only just balance B, and is insufficient to raise it. It must have been a force which raised B; that force must have been something more

FIG. 71.

than the weight of A, which was produced when the motion was checked. A was not stopped completely; it only lost some of its velocity, but it could not lose any velocity without being acted upon by a force. This force must have been applied by the rope by which A was held back, and the tension thus arising was sufficient to pull up B.

531. Let us remove the 14 lb. weight from B, and attach there a weight of 28 lbs., A remaining the same as before (14 lbs.). I raise A to the pulley; I allow it to fall. You observe that B, though double the weight of A, is again chucked up after the rope has become tight. We can only explain this by the supposition that the tension in the rope exerted in checking the motion of A is at least 28 lbs.

532. Finally, let us remove the 28 lbs. from B, put on 56 lbs., and perform the experiment again; you see that even the 56 lbs. is raised up several inches. Here a tension in the rope has been generated sufficient to overcome a weight four times as heavy as A. We have then, by the help of inertia, been able to produce a mechanical power, for a small force has overcome a greater.

533. After B is raised by the chuck to a certain height it descends again, if heavier than A, and raises A. The height to which B is raised is of course the same as the height through which A descends. You noticed that the height through which 28 lbs. was raised was considerably greater than that through which the 56 lbs. was raised. Hence we may draw the inference, that when A was deprived of its velocity while passing through a short space, it required to be opposed by a greater force than when it was gradually deprived of its velocity through a longer space. This is a most important point. Supposing I were to put a hundredweight at B, I have little doubt, if the rope were strong enough to bear the strain, that though A only weighs 14 lbs., B would yet be raised a little: here A would be deprived of its motion in a very short space, but the force required to arrest it would be very great.

534. It is clear that matters would not be much altered

if A were to be stopped by some force, exerted from below rather than above; in fact, we may conceive the rope omitted, and suppose A to be a hammer-head falling upon a nail in a piece of wood. The blow would force the nail to penetrate a small distance, and the entire velocity of A would have to be destroyed while moving through that small distance: consequently the force between the head of the nail and the hammer would be a very large one. This explains the effect of a blow.

535. In the case that we have supposed, the weight merely drops upon the nail: this is actually the principle of the hammer used in pile-driving machines. A pile is a large piece of timber, pointed and shod with iron at one end: this end is driven down into the ground. Piles are required for various purposes in engineering operations. They are often intended to support the foundations of buildings; they are therefore driven until the resistance with which the ground opposes their further entrance affords a guarantee that they shall be able to bear what is required.

536. The machine for driving piles consists essentially of a heavy mass of iron, which is raised to a height, and allowed to fall upon the pile. The resistance to be overcome depends upon the depth and nature of the soil: a pile may be driven two or three inches with each blow, but the less the distance the pile enters each time, the greater is the actual pressure with which the blow forces it downwards. In the ordinary hammer, the power of the arm imparts velocity to the hammer-head, in addition to that which is due to the fall; the effect produced is merely the same as if the hammer had fallen from a greater height.

537. Another point may be mentioned here. A nail will only enter a piece of wood when the nail and the wood are pressed together with sufficient force. The nail is urged by

the hammer. If the wood be lying on the ground, the reaction of the ground prevents the wood from getting away and the nail will enter. In other cases the element of *time* is all-important. If the wood be massive less force will make the nail penetrate than would suffice to move the wood quickly enough. If the wood be thin and unsupported, less force may be required to make it yield than to make the nail penetrate. The usual remedy is obvious. Hold a heavy mass close at the back of the wood. The nail will then enter because the augmented mass cannot now escape as rapidly as before.

THE STORING OF ENERGY.

538. Our study of the subject will be facilitated by some considerations founded on the principles of energy. In the experiment of Fig. 71 let A be 14 lbs., and B, on the ground, be 56 lbs. Since the rope is 15' long, A is 3' from the ground, and therefore 6' from the pulley. I raise A to the pulley, and, in doing so, expend $6 \times 14 = 84$ units of energy. Energy is never lost, and therefore I shall expect to recover this amount. I allow A to fall; when it has fallen 6', it is then precisely in the same condition as it was before being raised, except that it has a considerable velocity of descent. In fact, the 84 units of energy have been expended in giving velocity to A. B is then lifted to a maximum height x, in which $56 \times x$ units of energy have been consumed. At the instant when B is at the summit x, A must be at a distance of $6 + x$ feet from the pulley; hence the quantity of work performed by A is $14 \times (6 + x)$. But the work done by A must be equal to that done upon B, and therefore

$$14 (6 + x) = 56 x,$$

whence $x = 2$. If there were no loss by friction, B would therefore be raised 2'; but owing to friction, and doubtless also to the imperfect flexibility of the rope, the effect is not so great. We may regard the work done in raising A as so much energy stored up, and when A is allowed to fall, the energy is reproduced in a modified form.

539. Let us apply the principle of energy to the pile-driving engine to which we have referred (Art. 536); we shall then be able to find the magnitude of the force developed in producing the blow. Suppose the "monkey," that is the heavy hammer head, weighs 560 lbs. (a quarter of a ton). A couple of men raise this by means of a small winch to a height of 15'. It takes them a few minutes to do so; their energy is then saved up, and they have accumulated a store of $560 \times 15 = 8,400$ units. When the monkey falls upon the top of the pile it transfers thereto nearly the whole of the 8,400 units of energy, and this is expended in forcing the pile into the ground. Suppose the pile to enter one inch, the reaction of the pile upon the monkey must be so great that the number of units of energy consumed in one inch is 8,400. Hence this reaction must be $8,400 \times 12 = 100,800$ lbs. If the reaction did not reach this amount, the monkey could not be brought to rest in so short a distance. The reaction of the pile upon the monkey, and therefore the action of the monkey upon the pile, is about 45 tons. This is the actual pressure exerted.

540. If the soil which the pile is penetrating be more resisting than that which we have supposed,—for example, if the pile require a direct pressure of 100 tons to force it in,—the same monkey with the same fall would still be sufficient, but the pile would not be driven so far with each blow. The pressure required is 224,000 lbs.: this exerted over a space of $0'''\!\cdot\!45$ would be 8,400 units of energy; hence the pile would be driven $0''\!\cdot\!45$. The more the resistance, the less the penetration produced by each blow.

S

A pile intended to bear a very heavy load permanently must be driven until it enters but little with each blow.

541. We may compare the pile-driver with the mechanical powers in one respect, and contrast it in another. In each, we have machines which receive energy and restore it modified into a greater power exerted through a smaller distance; but while the mechanical powers restore the energy at one end of the machine, simultaneously with their reception of it at the other, the pile-driver is a reservoir for keeping energy which will restore it in the form wanted.

542. We have, then, a class of mechanical powers, of which a hammer may be taken as the type, which depend upon the storage of energy; the power of the arm is accumulated in the hammer throughout its descent, to be instantly transferred to the nail in the blow. Inertia is the property of matter which qualifies it for this purpose. Energy is developed by the explosion of gunpowder in a cannon. This energy is transferred to the ball, from which it is again in large part passed on to do work against the object which is struck. Here we see energy stored in a rapidly moving body, a case to which we shall presently return.

543. But energy can be stored in many ways; we might almost say that gunpowder is itself energy in a compact and storable form. The efforts which we make in forcing air into an air-cane are preserved as energy there stored to be reproduced in the discharge of a number of bullets. During the few seconds occupied in winding a watch, a small charge of energy is given to the spring which it expends economically over the next twenty-four hours. In using a bow my energy is stored up from the moment I begin to pull the string until I release the arrow.

544. Many machines in extensive use depend upon these

principles. In the clock or watch the demand for energy to sustain the motion is constant, while the supply is only occasional; in other cases the supply is constant, while the demand is only intermittent. We may mention an illustration of the latter. Suppose it be required *occasionally* to hoist heavy weights rapidly up to a height. If an engine sufficiently powerful to raise the weights be employed, the engine will be idle except when the weights are being raised; and if the machinery were to have much idle time, the waste of fuel in keeping up the fire during the intervals would often make the arrangement uneconomical. It would be a far better plan to have a smaller engine; and even though this were not able to raise the weight directly with sufficient speed, yet by keeping the engine continually working and storing up its energy, we might produce enough in the twenty-four hours to raise all the weights which it would be necessary to lift in the same time.

545. Let us suppose we want to raise slates from the bottom of a quarry to the surface. A large pulley is mounted at the top of the quarry, and over this a rope is passed: to each end of the rope a bucket is attached, so that when one of these is at the bottom the other is at the top, and their sizes and that of the pulley are so arranged that they pass each other with safety. A reservoir is established at the top of the quarry on a level with the pulley, and an engine is set to work constantly pumping up water from the bottom of the quarry into the reservoir. Each of the buckets is partly composed of a large tank, which can be quickly filled or emptied. The lower bucket is loaded with slates, and when ready for work, the man at the top fills the tank of the upper bucket with water: this accordingly becomes so heavy that it descends and raises the slates. When the heavier one reaches the bottom, the

water from its tank is let out into the lower reservoir, from which the engine pumps, and the slates are removed from the bucket which has been raised. All is then ready for a repetition of the same operation. If the slates be raised at intervals of ten minutes, the energy of the engine will be sufficient when in ten minutes' work it can pump up enough water to fill one tank; by the aid of this contrivance we are therefore able to accumulate for one effort the whole power of the engine for ten minutes.

THE FLY-WHEEL.

546. One of the best means of storing energy is by setting a heavy body in rapid motion. This has already been referred to in the case of the cannon-ball. In order to render this method practically available for the purposes of machinery, the heavy body we use is a fly-wheel, and the rapid motion imparted to it is that of rotation about its axis. A very large amount of energy can by this means be stored in a manageable form.

547. We shall illustrate the principle by the apparatus of Fig. 72. This represents an iron fly-wheel B: its diameter is 18", and its weight is 26 lbs.; the fly is carried upon a shaft (A) of wrought iron $\frac{3}{4}$" in diameter. We shall store up a quantity of energy in this wheel, by setting it in rapid motion, and then we shall see how we can recover from it the energy we have imparted.

548. A rope is coiled around the shaft; by pulling this rope the wheel is made to turn round: thus the rope is the medium by which my energy shall be imparted to the wheel. To measure the operation accurately, I attach the rope to the hook of the spring balance (Fig. 9); and by taking the ring of the balance in my hand, I learn from the index the amount of the force I am exerting. I find that when I walk

backwards as quickly as is convenient, pulling the rope all the time, the scale shows a tension of about 50 lbs. To set the wheel rapidly in motion, I pull about 20' of rope from the axle, so that I have imparted to the wheel somewhere about $50 \times 20 = 1,000$ units of energy. The rope is fastened to the shaft, so that, after it has been all unwound, the wheel now rapidly rotating winds it in. By measuring the time in which the wheel made a certain number of coils of the rope around the shaft, I find that it makes about 600 revolutions per minute.

Fig. 72.

549. Let us see how the stored-up energy can be exhibited. A piece of pine $24'' \times 1'' \times 1''$ of which both ends are supported, requires a force of 300 lbs. applied to its centre to produce fracture (See p. 190). I arrange such a piece of pine near the wheel. As the shaft is winding in the rope, a tremendous chuck would be given to anything which tried to stop the

motion. If I tied the end of the rope to the piece of pine, the chuck would break the rope; therefore I have fastened one end of a 10' length of chain to the rope, and the other has been tied round the middle of the wooden bar. The wheel first winds in the rope, then the chain takes a few turns before it tightens, and crack goes the rod of pine. The wheel had no choice; it must either stop or break the rod: but nature forbids it to be stopped, unless by a great force, which the rod was not strong enough to apply. Here I never exerted a force greater than 50 lbs. in setting the wheel in motion. The wheel stored up and modified my energy so as to produce a force of 300 lbs., which had, however, only to be exerted over a very small distance.

550. But we may show the experiment in another way, which is that represented in the figure (72). We see the chain is there attached to two 56 lb. weights. The mode of proceeding is that already described. The rope is first wound round the shaft, then by pulling the rope the wheel is made to revolve; the wheel then begins to wind in the rope again, and when the chain tightens the two 56 lbs. are raised up to a height of 3 or 4 feet. Here, again, the energy has been stored and recovered. But though the fly-wheel will thus preserve energy, it does so at some cost: the store is continually being frittered away by friction and the resistance of the air; in fact, the energy would altogether disappear in a little time, and the wheel would come to rest; it is therefore economical to make the wheel yield up what it has received as soon as possible.

551. These principles are illustrated by the function of the fly-wheel in a steam-engine. The pressure of the steam upon the piston varies according to the different parts of the stroke: and the fly-wheel obviates the inconvenience which would arise from such irregularity. Its great inertia makes its velocity but

little augmented by the exuberant action of the piston when the pressure is greatest, while it also sustains the motion when the piston is giving no assistance. The fly-wheel is a vast reservoir into which the engine pours its energy, sudden floods alternating with droughts; but these succeed each other so rapidly, and the area of the reservoir is so vast, that its level remains sensibly uniform, and the supplies sent out to the consumers are regular and unvaried. The consumers of the energy stored in the fly-wheel of an engine are the machines in the mill; they are supplied by shafts which traverse the building, conveying, by their rotation, the energy originally condensed within the coal from which combustion has set it free.

THE PUNCHING MACHINE.

552. When energy has been stored in a fly-wheel, it can be withdrawn either as a small force acting over a great distance, or as a large force over a small distance. In the latter case the fly-wheel acts as a mechanical power, and in this form it is used in the very important machine to be next described. A model of the punching machine is shown in Fig. 73.

The punching machine is usually worked by a steam-engine, a handle will move our small model. The handle turns a shaft on which the fly-wheel F is mounted. On the shaft is a small pinion D of 40 teeth: this works into a large wheel E of 200 teeth, so that, when the fly and the pinion have turned round 5 times, E will have turned round once. C is a circular piece of wood called a cam, which has a hole bored through it, between the centre and circumference; by means of this hole, the cam is mounted on the same axle as E, to which it is rigidly fastened, so that the two must revolve together. A is a lever of the first order, whose fulcrum is at

A: the remote end of this lever rests upon the cam C; the other end B contains the punch. As the wheel E revolves it carries with it the cam: this raises the lever and forces the punch down a hole in a die, into which it fits exactly. The metal to be operated on is placed under the punch before it is depressed by the cam, and the pressure drives the punch through, cutting out a cylindrical piece of metal from the plate: this model will, as you see, punch through ordinary tin.

FIG. 73.

553. Let us examine the mode of action. The machine being made to rotate rapidly, the punch is depressed once for every 5 revolutions of the fly; the resistance which the metal opposes to being punched is no doubt very great, but the lever acts at a twelve-fold advantage. When the punch comes down on the surface of the metal, one of three things must happen: either the motion must stop suddenly, or the machine must be strained and injured, or the metal must be punched. But the motion cannot be stopped suddenly, because, before this could happen, an infinite force

would be developed, which must make something yield. If therefore we make the structure sufficiently massive to prevent yielding, the metal must be punched. Such machines are necessarily built strong enough to make the punching of the metal easier than breaking the machine.

554. We shall be able to calculate, from what we have already seen in Art. 248, the magnitude of the force required for punching. We there learned that about 22·5 tons of pressure was necessary to shear a bar of iron one square inch in section. Punching is so far analogous to shearing that in each case a certain area of surface has to be cut; the area in punching is measured by the cylinder of iron which is removed.

555. Suppose it be required to punch a hole 0"·5 in diameter through a plate 0"·8 thick, the area of iron that has to be cut across is $\frac{22}{7} \times \frac{1}{2} \times \frac{4}{5} = 1·26$ square inches: and as 22·5 tons per square inch are required for shearing, this hole will require $22·5 \times 1·26 = 28·4$ tons. A force of this amount must therefore be exerted upon the punch: which will require from the cam a force of more than 2 tons upon its end of the lever. Though the iron must be pierced to a depth of 0'"·8, yet it is obvious that almost immediately after the punch has penetrated the surface of the iron, the cylinder must be entirely cut and begin to emerge from the other side of the plate. We shall certainly be correct in supposing that the punching is completed before the punch has penetrated to a depth of 0"·2, and that for not more than this distance has the great pressure of 28 tons been exerted; for a small pressure is afterwards sufficient to overcome the friction which opposes the motion of the cylinder of iron. Hence, though so great a pressure has been required, yet the number of units of energy consumed is not very large; it is $\frac{1}{60} \times 2240 \times 28·4 = 1062$.

The energy actually required to punch a hole of half an inch diameter through a plate eight-tenths of an inch thick is therefore less than that which would be expended in raising 1 cwt. to a height of ten feet.

556. The fly-wheel may be likened to the reservoir in Art 545. The time that is actually occupied in the punching is extremely small, and the sudden expenditure of 1062 units is gradually reimbursed by the engine. If the rotating fly-wheel contain 50,000 units of energy, the abstraction of 1362 units will not perceptibly affect its velocity. There is therefore an advantage in having a heavy fly sustained at a high speed for the working of a punching machine.

LECTURE XVII.

CIRCULAR MOTION.

The Nature of Circular Motion.—Circular motion in Liquids.—The Applications of Circular Motion.—The Permanent Axes.

THE NATURE OF CIRCULAR MOTION.

557. To compel a body to swerve from motion in a straight line force must be exercised. In this chapter we shall study the comparatively simple case of a body revolving in a circle.

558. When a body moves round uniformly in a circle force must be continuously applied, and the first question for us to examine is, as to the *direction* of that force. We have to demonstrate the important fact, that it constantly tends towards the centre.

559. The direction of the force can be exhibited by actual experiment, and its magnitude will be at the same time clearly indicated by the extent to which a spring is stretched. The apparatus we use is shown in Fig. 74. The essential parts of the machine consist of two balls

A, B, each 2″ in diameter: these are thin hollow spheres of silvered brass. The balls are supported on arms P A, Q B, which are attached to a piece of wood, P Q, capable of turning on a socket at C. The arm A P is rigidly fixed to P Q; the other arm, B Q, is capable of turning round a pin at Q. An india-rubber door-spring is shown at F;

Fig. 74.

one end of this is secured to P Q, the other end to the movable arm, Q B. If the arm Q B be turned so as to move B away from C, the spring F must be stretched.

A small toothed wheel is mounted on the same socket as C; this is behind P Q, and is therefore not seen in the figure: the whole is made to revolve rapidly by the large wheel E, which is turned by the handle D.

560. The room being darkened, a beam from the limelight is allowed to fall on the apparatus: the reflections of the light are seen in the two silvered balls as two bright points. When D is turned, the balls move round rapidly, and you see the points of light reflected from them describe circles. The ball B when at rest is 4" from C, while A is 8" from C; hence the circle described by B is smaller than that described by A. The appearance presented is that of two concentric luminous circles. As the speed increases, the inner circle enlarges till the two circles blend into one. By increasing the speed still more, you see the circle whose diameter is enlarging actually exceeding the fixed circle, and its size continues to increase until the highest velocity which it is safe to employ has been communicated to the machine.

561. What is the explanation of this? The arm A is fixed and the distance A C cannot alter, hence A describes the fixed circle. B, on the other hand, is not fixed; it can recede from C, and we find that the quicker the speed the further it recedes. The larger the circle described by B the more is the spring stretched, and the greater is the force with which B is attracted towards the centre. This experiment proves that the force necessary to retain a body in a circular path must be increased when the speed is increased.

562. Thus we see that uniform motion of a body in a circle can only be produced by an uniform force directed to the centre.

If the motion, even though circular, have variable speed the law of the force is not so simple.

563. We can measure the magnitude of this force by the same apparatus. The ball B weighs 0·1 lb. I find that I

must pull it with a force of 3 lbs. in order to draw it to a distance of 8" from C; that is, to the same distance as A is from C. Hence, when the diameters of the circles in which the balls move are equal, the central force must be 3 lbs.; that is, it must be nearly thirty times as great as gravity.

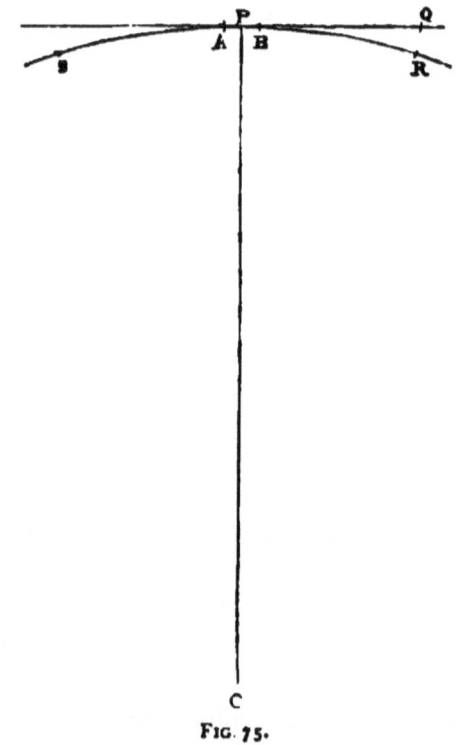

Fig. 75.

564. The necessity for the central force is thus shown: Let us conceive a weight attached to a string to be swung round in a circle, a portion of which is shown in Fig. 75.

Suppose the weight be at S and moving towards P, and let a tangent to the circle be drawn at P. Take two points

XVII.] ACTION OF CIRCULAR MOTION. 271

on the circle, A and B, very near P; the small arc A B does not differ perceptibly from the part A B on the tangent line; hence, when the particle arrives at A, it is a matter of indifference whether it travels in the arc A B, or along the line A B. Let us suppose it to move along the line. By the first law of motion, a particle moving in the line A B would continue to do so; hence, if the particle be allowed, it will move on to Q: but the particle is not allowed to move to Q; it is found at R. Hence it must have been withdrawn by some force.

565. This force is supplied by the string to which the weight is attached. The incessant change from the rectilinear motion of the weight requires a constantly applied force, and this is always directed to the centre. Should the string be released, the body flies off in the direction of the tangent to the circle at the point which the body occupied at the instant of release.

566. The central force increases in proportion to the square of the velocity. If I double the speed with which the weight is whirled round in the circle, I quadruple the force which the string must exert on the body. If the speed be trebled, the force is increased ninefold, and so on. When the speeds with which two equal masses are revolving in two circles are equal, the central force in the smaller circle is greater than that of the larger circle, in the proportion of the radius of the larger circle to that of the smaller.

THE ACTION OF CIRCULAR MOTION UPON LIQUIDS.

567. I have here a small bucket nearly filled with water: to the handle a piece of string is attached. If I whirl the

bucket round in a vertical plane sufficiently fast, you see no water escapes, although the bucket is turned upside down once in every revolution. This is because the water has not *time* to fall out during such a brief interval. A body would not fall half an inch from rest in the twentieth of a second.

568. The action of circular motion upon liquids is illustrated by the experiment which is represented in Fig. 76.

FIG. 76.

A glass beaker about half full of water is mounted so that it can be spun round rapidly. The motion is given by means of a large wheel turned by a handle, as shown in the figure. When the rotation commences, the water is seen to rise up against the glass sides and form a hollow in the centre.

569. In order to demonstrate this clearly, I turn upon the vessel a beam from the lime-light. I have previously

dissolved a little quinine in the water. The light from the lamp is transmitted through a piece of dense blue glass. When the light thus coloured falls on the water, the presence of the quinine makes the entire liquid glow with a bluish hue. This remarkable property of quinine, which is known as fluorescence, enables you to see distinctly the hollow form caused by the rotation.

570. You observe that as the speed becomes greater the depth of the hollow increases, and that if I turn the wheel sufficiently fast the water is actually driven out of the glass. The shape of the curve which the water assumes is that which would be produced by the revolution of a *parabola* about its axis.

571. The explanation is simple. As soon as the glass begins to revolve, the friction of its sides speedily imparts a revolving motion to the water; but in this case there is nothing to keep the particles near the centre like the string in the revolving weight, so the liquid rises at the sides of the glass.

572. But you may ask why *all* the particles of the water should not go to the circumference, and thus line the inside of the glass with a hollow cylinder of water instead of the parabola. Such an arrangement could not exist in a liquid acted on by gravity. The lower parts of the cylinder must bear the pressure of the water above, and therefore have more tendency to flatten out than the upper portions. This tendency could not be overcome by any consequences of the movement, for such must be alike on all parts at the same distance from the axis.

573. A very beautiful experiment was devised by Plateau for the purpose of studying the revolution of a liquid removed from the action of gravity.

The apparatus employed is represented in Fig. 77. A

T

glass vessel 9″ cube is filled with a mixture of alcohol and water. The relative quantities ought to be so proportioned that the fluid has the same specific gravity as olive oil, which is heavier than alcohol and lighter than water. In practice, however, it is found so difficult to adjust the composition exactly that the best plan is to make two alcoholic mixtures so that olive oil will just float on one of them, and just

FIG. 77.

sink in the other. The lower half of the glass is to be filled with the denser mixture, and the upper half with the lighter. If, then, the oil be carefully introduced with a funnel it will form a beautiful sphere in the middle of the vessel, as shown in the figure. We thus see that a liquid mass freed from the action of terrestrial gravity, forms a sphere by the mutual attraction of its particles.

Through the liquid a vertical spindle passes. On this there is a small disk at the middle of its length, about which the sphere of oil arranges itself symmetrically. To the end of the spindle a handle is attached. When the handle is turned round slowly, the friction of the disk and spindle communicates a motion of rotation to the sphere of oil. We have thus a liquid spheroidal mass endowed with a movement of rotation; and we can study the effect of the motion upon its form. We first see the sphere flatten down at its poles, and bulge at its equator. In order to show the phenomenon to those who may not be near to the table, the sphere can be projected on the screen by the help of the lime-light lamp and a lens. It first appears as a yellow circle, and then, as the rotation begins, the circle gradually transforms into an ellipse. But a very remarkable modification takes place when the handle is turned somewhat rapidly. The ellipsoid gradually flattens down until, when a certain velocity has been attained, the surface actually becomes indented at the poles, and flies from the axis altogether. Ultimately the liquid assumes the form of a beautiful ring, and the appearance on the screen is shown in Fig. 78.

574. The explanation of the development of the ring involves some additional principles: as the sphere of oil spins round in the liquid, its surface is retarded by friction; so that when the velocity attains a certain amount, the internal portions of the sphere, which are in the neighbourhood of the spindle, are driven from the centre into the outer portions, but the full account of the phenomenon cannot be given here.

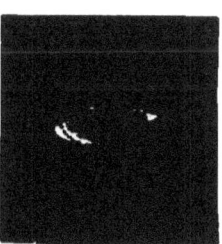

FIG. 78.

575. The earth was, we believe, originally in a fluid

condition. It had then, as it has now, a diurnal rotation, and one of the consequences of this rotation has been to cause the form to be slightly protuberant at the equator, just as we have seen the sphere of oil to bulge out under similar conditions.

576. Bodies lying on the earth are whirled around in a great circle every day. Hence, if there were not some force drawing them to the centre, they would fly off at a tangent. A part of the earth's attraction goes for this purpose, and the remainder, which is the apparent weight, is thus diminished by a quantity increasing from the pole to the equator (Art. 86).

THE APPLICATIONS OF CIRCULAR MOTION.

577. These principles have many applications in the mechanical arts; we shall mention two of them. The first is to the governor-balls of a steam-engine; the second is to the process of sugar-refining.

An engine which turns a number of machines in a factory should work uniformly. Irregularities of motion may be productive of loss and various inconveniences. An engine would work irregularly either from variation in the production of steam, or from the demands upon the power being lessened or increased. Even if the first of these sources of irregularity could be avoided by care, it is clear that the second could not. Some machines in the mill are occasionally stopped, others occasionally set in motion, and the engine generally tends to go faster the less it has to do. It is therefore necessary to provide means by which the speed shall be restrained within narrow limits, and it is obviously desirable that the contrivance used for this purpose should be self-acting. We must, therefore, have some arrangement which shall admit more steam to the cylinder when

XVII.] APPLICATIONS OF CIRCULAR MOTION. 277

the engine is moving too slowly, and less steam when it is moving too quickly. The valve which is to regulate this must, be worked by some agent which depends upon the velocity of the engine; this at once points to circular motion because the force acting on the revolving body depends upon its velocity. Such was the train of reasoning which led to the happy invention of the governor-balls: these are shown in Fig. 79.

A B is a vertical spindle which is turned by the engine. P P is a piece firmly attached to the spindle and turning with it. P W, P W are arms terminating in weights W W; these are balls of iron, generally very massive: the arms are free to turn round pins at P P. At Q Q links are placed, attached to another piece R R, which is free to slide up and down the spindle. When A B rotates, W and W are carried round, and therefore fly outwards from the spindle; to do this they must evidently pull the piece R R up the shaft. We can easily imagine an arrangement by which R R shall be made to shut or open the steam-valve according as it ascends or descends. The problem is then solved, for if the engine begin to go too rapidly, the balls fly out further just as they did in Fig. 74: this movement raises the piece R R, which diminishes the supply of steam,

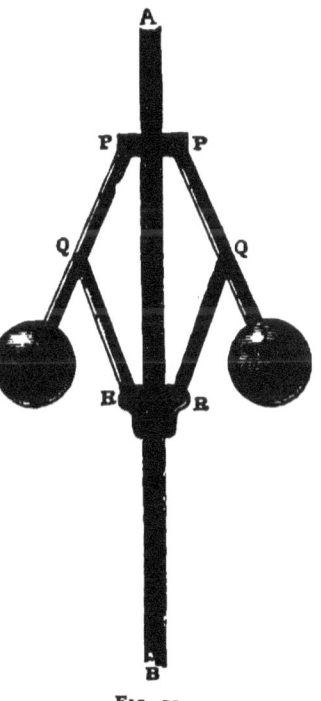

FIG. 79.

and consequently checks the speed. On the other hand, when the engine works too slowly, the balls fall in towards the spindle, the piece R R descends, the valve is opened, and a greater supply of steam is admitted. The objection to this governor is that though it moderates, it does not completely check irregularity. There are other governors occasionally employed which depend also on circular motion; some of these are more sensitive than the governor-balls; but they are elaborate machines, only to be employed under exceptional circumstances.

578. The application of circular motion to sugar-refining is a very beautiful invention. To explain it I must briefly describe the process of refining.

The raw sugar is dissolved in water, and the solution is purified by straining and by filtration through animal charcoal. The syrup is then boiled. In order to preserve the colour of the sugar, and to prevent loss, this boiling is conducted *in vacuo*, as by this means the temperature required is much less than would be necessary with the ordinary atmospheric pressure.

The evaporation having been completed, crystals of sugar form throughout the mass of syrup. To separate these crystals from the liquor which surrounds them, the aid of circular motion force is called in. A mass of the mixture is placed in a large iron tub, the sides of which are perforated with small holes. The tub is then made to rotate with prodigious velocity; its contents instantly fly off to the circumference, the liquid portions find an exit through the perforations in the sides, but the crystals are left behind. A little clear syrup is then sprinkled over the sugar while still rotating: this washes from the crystals the last traces of the coloured liquid, and passes out through the holes; when the motion ceases, the inside of the tub contains a layer of perfectly pure white sugar, some inches in thickness, ready for the market.

579. Circular motion is peculiarly fitted for this purpose; each particle of liquid strives to get as far away from the axis as possible. The action on the sugar is very different from what it would have been had the mass been subjected to pressure by a screw-press or similar contrivance; the particles immediately acted on would then have to transmit the pressure to those within; and the consequence would be that while the crystals of sugar on the outside would be crushed and destroyed, the water would only be very imperfectly driven from the interior: for it could lurk in the interstices of the sugar, which remain notwithstanding the pressure.

580. But with circular motion the water must go, not because it is pushed by the crystals, but because of its own inertia; and it can be perfectly expelled by a velocity of rotation less than that which would be necessary to produce sufficient pressure to make the crystals injure each other.

THE PERMANENT AXES.

581. There are some curious properties of circular motion which remain to be considered. These we shall investigate by means of the apparatus of Fig. 80. This consists of a pair of wheels B C, by which a considerable velocity can be given to a horizontal shaft. This shaft is connected by a pair of bevelled wheels D with a vertical spindle F. The machine is worked by a handle A, and the object to be experimented upon is suspended from the spindle.

582. I first take a disk of wood 18" in diameter; a hole is bored in the margin of this disk; through this hole a

rope is fastened, by means of which the disk is suspended from the spindle. The disk hangs of course in a vertical plane.

583. I now begin to turn the handle round gently,

Fig. 80

and you see the disk begins to rotate about the vertical diameter; but, as the speed increases, the motion becomes a little unsteady; and finally, when I turn the handle very rapidly, the disk springs up into a horizontal plane, and

XVII.] THE PERMANENT AXES. 281

you see it like the surface of a small table: the rope swings round and round in a cone, so rapidly that it is hardly seen.

584. We may repeat the experiment in a different manner. I take a piece of iron chain about 2' long, G; I pass the rope through the two last links of its extremities, and suspend the rope from the spindle. When I commence to turn the handle, you see the chain gradually opens out

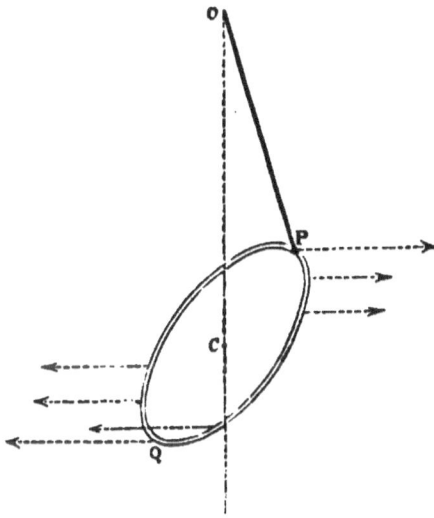

FIG. 81.

into a loop H; and as the speed increases, the loop becomes a complete ring. Still increasing the speed, I find the ring becomes unsteady, till finally it rises into a horizontal plane. The ring of chain in the horizontal plane is shown at I. When the motion is further increased, the ring swings about violently, and so I cease turning the handle.

585. The principles already enunciated will explain these

remarkable results; we shall only describe that of the chain, as the same explanation will include that of the disk of wood. We shall begin with the chain hanging vertically from the spindle: the moment rotation commences, the chain begins to spin about a vertical axis; the parts of the chain fly outwards from this axis just as the ball flies outwards in Fig. 74; this is the cause of the looped form H which the chain assumes. As the speed is increased the loop gradually opens more and more, just as the diameter of the circle Fig. 74 increases with the velocity. But we have also to inquire into the cause of the remarkable change of position which the ring undergoes; instead of continuing to rotate about a vertical diameter, it comes into a horizontal plane. This will be easily understood with the help of Fig. 81. Let O P represent the rope attached to the ring, and O C be the vertical axis. Suppose the ring to be spinning about the axis O C, when O C was a diameter; if then, from any cause, the ring be slightly displaced, we can show that the circular motion will tend to drive the ring further from the vertical plane, and force it into the horizontal plane. Let the ring be in the position represented in the figure; then, since it revolves about the vertical line O C, the tendency of P and Q is to move outwards in the directions of the arrows, thus evidently tending to bring the ring into the horizontal plane.

586. In Art. 103 we have explained what is meant by stable and unstable equilibrium; we have here found a precisely analogous phenomenon in motion. The rotation of the ring about its diameter is unstable, for the minutest deviation of the ring from this position is fatal; circumstances immediately combine to augment the deviation more and more, until finally the ring is raised into the horizonal plane. Once in the horizontal plane, the motion there is

stable, for if the ring be displaced the tendency is to restore it to the horizontal.

587. The ring, when in a horizontal plane, rotates permanently about the vertical axis through its centre; this axis is called permanent, to distinguish it from all other directions, as being the only axis about which the motion is stable.

588. We may show another experiment with the chain : if instead of passing the rope through the links at its ends, I pass the rope through the centre of the chain, and allow the ends of the chain to hang downwards. I now turn the handle; instantly the parts of the chain fly outwards in a curved form; and by increasing the velocity, the parts of the chain at length come to lie almost in a straight line.

LECTURE XVIII.

THE SIMPLE PENDULUM.

Introduction.—The Circular Pendulum.—Law connecting the Time of Vibration with the Length.—The Force of Gravity determined by the Pendulum.—The Cycloid.

INTRODUCTION.

589. IF a weight be attached to a piece of string, the other end of which hangs from a fixed point, we have what is called a simple pendulum. The pendulum is of the utmost importance in science, as well as for its practical applications as a time-keeper. In this lecture and the next we shall treat of its general properties; and the last will be devoted to the practical applications. We shall commence with the simple pendulum, as already defined, and prove, by experiment, the remarkable property which was discovered by Galileo. The simple pendulum is often called the circular pendulum.

THE CIRCULAR PENDULUM.

590. We first experiment with a pendulum on a large scale. Our lecture theatre is 32 feet high, and there is a

LECT. XVIII.] THE CIRCULAR PENDULUM. 285

wire suspended from the ceiling 27' long; to the end of this a ball of cast iron weighing 25 lbs. is attached. This wire when at rest hangs vertically in the direction o c (Fig. 82).

I draw the ball from its position of rest to A; when released, it slowly returns to C, its original position; it then moves on the other side to B, and back again to my hand at A. The ball—or to speak more precisely, the centre of the ball— moves in a circle, the centre being the point o in the ceiling from which the wire is suspended.

591. What causes the motion of the pendulum when the weight is released? It is the force of gravity; for by moving the ball to A I raise it a little, and therefore, when I release it gravity compels it to return to C it being the only manner in which the mode of suspension will allow it to fall. But when it has reached its original position at C, why does it continue its motion? — for gravity must be acting against the ball during the journey from C to B. The first law of motion explains this. (Art 485). In travelling from A to C the ball has acquired a certain velocity, hence it has a tendency to go on, and only by the time it has arrived at B will gravity have arrested the velocity, and begin to make it descend.

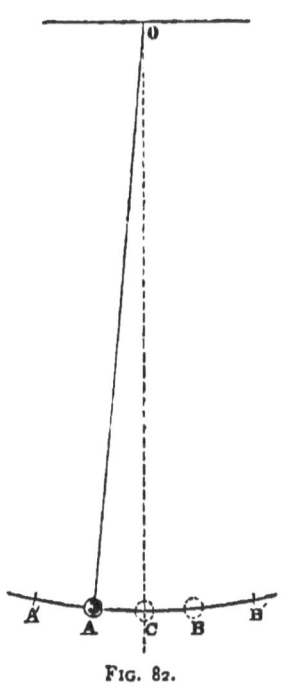

FIG. 82.

592. You see, the ball continues moving to and fro— oscillating, as it is called—for a long time. The fact is that

it would oscillate for ever, were it not for the resistance of the air, and for some loss of energy at the point of suspension.

593. By the time of an oscillation is meant the time of going from A to B, but not back again. The time of our long pendulum is nearly three seconds.

594. With reference to the time of oscillation Galileo made a great discovery. He found that whether the pendulum were swinging through the arc A B, or whether it had been brought to the point A', and was thus describing the arc A' B', the time of oscillation remained nearly the same. The arc through which the pendulum oscillates is called its amplitude, so that we may enunciate this truth by saying that *the time of oscillation is nearly independent of the amplitude*. The means by which Galileo proved this would hardly be adopted in modern days. He allowed a pendulum to perform a certain number of vibrations, say 100, through the arc A B, and he counted his pulse during the time; he then counted the number of pulsations while the pendulum vibrated 100 times in the arc A' B', and he found the number of pulsations in the two cases to be equal. Assuming, what is probably true, that Galileo's pulse remained uniform throughout the experiment, this result showed that the pendulum took the same time to perform 100 vibrations, whether it swung through the arc A B, or through the arc A' B'. This discovery it was which first suggested the employment of the pendulum as a means of keeping time.

595. We shall adopt a different method to show that the time does not depend upon the amplitude. I have here an arrangement which is represented in Fig. 83. It consists of two pendulums A D and B C, each 12' long, and suspended from two points A B, about 1' apart, in the same

XVIII.] THE CIRCULAR PENDULUM. 287

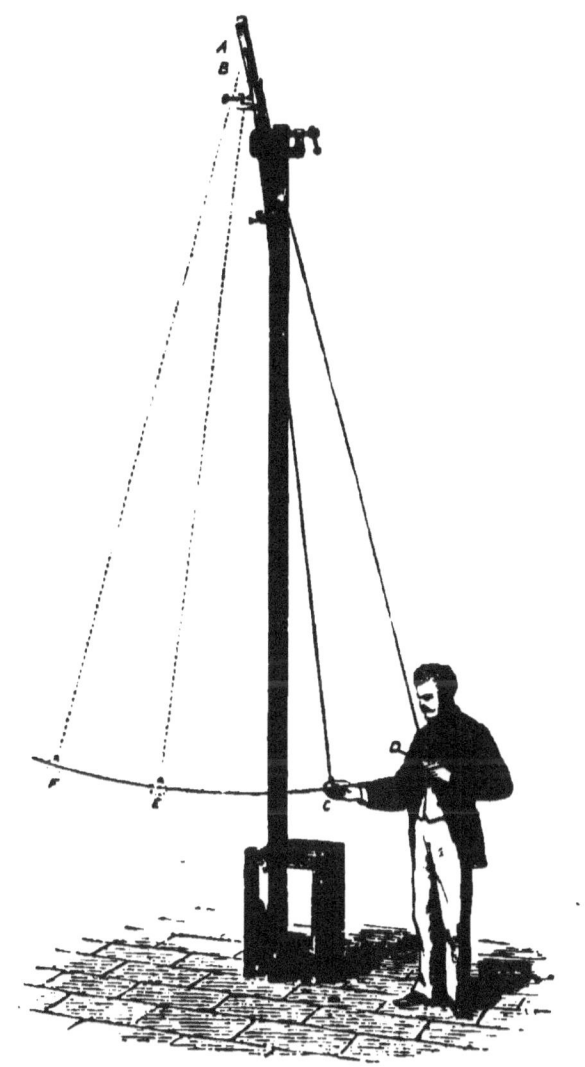

FIG. 83.

horizontal line. Each of these pendulums carries a weight of the same size: they are in fact identical.

596. I take one of the balls in each hand. If I withdraw each of them from its position of rest through equal distances and then release them, both balls return to my hands at the same instant. This might have been expected from the identity of the circumstances.

597. I next withdraw the weight C in my right hand to a distance of 1', and the weight D in my left hand to a distance of 2', and release them simultaneously. What happens? I keep my hands steadily in the same position, and I find that the two weights return to them at the same instant. Hence, though one of the weights moved through an amplitude of 2' (C E) while the other moved through an amplitude of 4' (D F), the times occupied by each in making two oscillations are identical. If I draw the right-hand ball away 3', while I draw the left hand only 1' from their respective positions of rest, I still observe the same result.

598. In two oscillations we can see no effect on the time produced by the amplitude, and we are correct in saying that, when the amplitude is only a small fraction of the length of the pendulum, its effect is inappreciable. But if the amplitude of one pendulum were very large, we should find that its time of oscillation is slightly greater than that of the other, though to detect the difference would require a delicate test. One consequence of what is here remarked will be noticed at a later page. (Art 655.)

599. We next inquire whether the weight which is attached to the pendulum has any influence upon the time of vibration. Using the 12' pendulums of Fig. 83, I place a weight of 12 lbs. on one hook and one of 6 lbs. on the other. I withdraw one in each hand; I release them; they return to my hand at the same moment. Whether I withdraw the weights through long arcs or short arcs, equal or unequal, they invariably return together, and both therefore

have the same time of vibration. With other iron weights the same law is confirmed, and hence we learn that, besides being independent of the amplitude, the time of vibration is also independent of the weight.

600. Finally, let us see if the *material* of the pendulum can influence its time of vibration. I place a ball of wood on one wire and a ball of iron on the other; I swing them as before : the vibrations are still performed in equal times. A ball of lead is found to swing in the same time as a ball of brass, and both in the same time as a ball of iron or of wood.

601. In this we may be reminded of the experiments on gravity (Art. 491), where we showed that all bodies fall to the ground in equal times, whatever be their sizes or their materials. From both cases the inference is drawn that the force of gravity upon different bodies is proportional to their masses, though the bodies be made of various substances. It was indeed by means of experiments with the pendulum that Newton proved that gravity had this property, which is one of the most remarkable truths in nature.

LAW CONNECTING THE TIME OF VIBRATION WITH THE LENGTH.

602. We have seen that the time of vibration of a pendulum depends neither upon its amplitude, material, nor weight; we have now to learn on what the time *does* depend. It depends upon the *length* of the pendulum. The shorter a pendulum the less is its time of vibration. We shall find by experiment the relation between the time and the length of the cord by which the weight is suspended.

603. I have here (Fig. 84) two pendulums A D, B C, one

of which is 12′ long and the other 3′; they are mounted side by side, and the weights are at the same distance

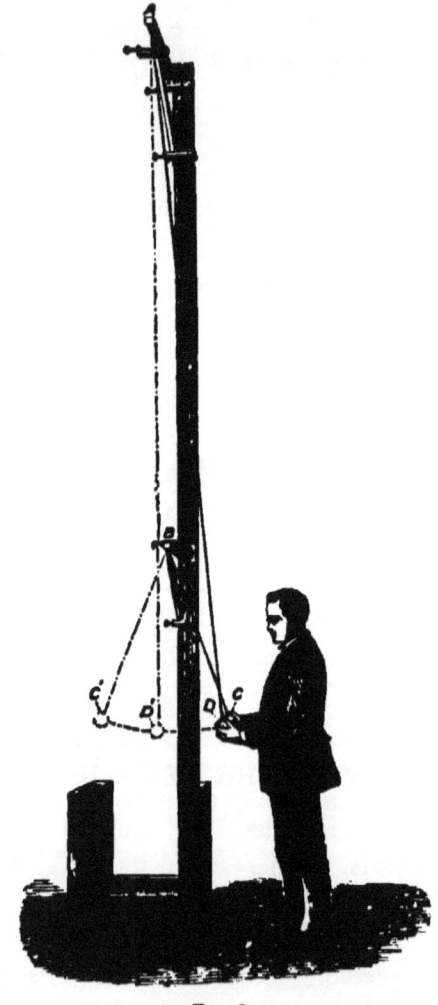

FIG. 84.

from the floor. I take one of the weights in each hand, and withdraw them to the same distance from the position of rest. I release the balls simultaneously; C moves off rapidly, arrives at the end C' while D has only reached D', and returns to my hand just as D has completed one oscillation. I do not seize C: it goes off again, only to return at the same moment when D reaches my hand. Thus C has performed four oscillations while D has made no more than two. This proves that when one of two pendulums is a quarter the length of the other, the time of vibration of the shorter one is half that of the other.

604. We shall repeat the experiment with another pendulum 27' long, suspended from the ceiling, and compare its vibrations with those of a pendulum 3' long. I draw the weights to one side and release them as before; and you see that the small pendulum returns twice to my hand while the long pendulum is still absent; but that, keeping my hands steadily in the same place throughout the experiment, the long pendulum at last returns simultaneously with the third arrival of the short one. Hence we learn that a pendulum 27' long takes three times as much time for a single vibration as a 3' pendulum.

605. The lengths of the three pendulums on which we have experimented (27', 12', 3'), are in the proportions of the numbers 9, 4, 1; and the times of the oscillations are proportional to 3, 2, 1 : hence we learn that *the period of oscillation of a pendulum is proportional to the square root of its length.*

606. But the time of vibration must also depend upon gravity; for it is only owing to gravity that the pendulum vibrates at all. It is evident that, if gravity were increased, all bodies would fall to the earth more than 16' in the first second. The effect on the pendulum would be to draw the ball more quickly from D to D' (Fig. 84), and thus the time of vibration would be diminished.

It is found by calculation, and the result is confirmed by experiment, that the time of vibration is represented by the expression,

$$3\cdot1416 \sqrt{\frac{\text{Length}}{\text{Force of gravity}}}.$$

607. The accurate value of the force of gravity in London is $32\cdot1908$, so that the time of vibration of a pendulum there is $0\cdot5537 \sqrt{\text{length}}$: the length of the seconds pendulum is $3''\cdot2616$.

THE FORCE OF GRAVITY DETERMINED BY THE PENDULUM.

608. The pendulum affords the proper means of measuring the force of gravity at any place on the earth. We have seen that the time of vibration can be expressed in terms of the length and the force of gravity; so conversely, when the length and the time of vibration are known, the force of gravity can be determined and the expression for it is—

$$\text{Length} \times \left(\frac{3\cdot1416}{\text{Time}}\right)^2.$$

609. It is impossible to observe the time of a single vibration with the necessary degree of accuracy; but supposing we consider a large number of vibrations, say 100, and find the time taken to perform them, we shall then learn the time of one oscillation by dividing the entire period by 100. The amplitudes of the oscillations may diminish, but they are still performed in the same time; and hence, if we are sure that we have not made a mistake of more than one second

in the whole time, there cannot be an error of more than 0·01 second, in the time of one oscillation. By taking a still larger number the time may be determined with the utmost precision, so that this part of the inquiry presents little difficulty.

610. But the length of the pendulum has also to be ascertained, and this is a rather baffling problem. The ideal pendulum whose length is required, is supposed to be composed of a very fine, perfectly flexible cord, at the end of which a particle without appreciable size is attached; but this is very different from the pendulum which we must employ. We are not sure of the exact position of the point of suspension, and, even if we had a perfect sphere for the weight, the distance between its centre and the point of suspension is not the same thing as the length of the simple pendulum that would vibrate in the same time. Owing to these circumstances, the measurement of the pendulum is embarrassed by considerable difficulties, which have however been overcome by ingenious contrivances to be described in the next chapter.

611. But we shall perform, in a very simple way, an experiment for determining the force of gravity. I have here a silken thread which is fastened by being clamped between two pieces of wood. A cast-iron ball $2''·54$ in diameter is suspended by this piece of silk. The distance from the point of suspension of the silk to the ball is $24''·07$, as well as it can be measured.

The length of the ideal pendulum which would vibrate isochronously with this pendulum is $25'''37$, being about $0'''03$ greater than the distance from the point of suspension to the centre of the sphere.

12. The length having been ascertained, the next element to be determined is the time of vibration. For this purpose

I use a stop-watch, which can be started or stopped instantaneously by touching a little stud : this watch will indicate time accurately to one-fifth of a second. It is necessary that the pendulum should swing in a small arc, as otherwise the oscillations are not strictly isochronous. Quite sufficient amplitude is obtained by allowing the ball to rotate to and fro through a few tenths of an inch.

613. In order to observe the movement easily, I have mounted a little telescope, through which I can view the top of the ball. In the eye-piece of the telescope a vertical wire is fastened, and I count each vibration just as the silken thread passes the vertical wire. Taking my seat with the stop-watch in my hand, I write down the position of the hands of the stop-watch, and then look through the telescope. I see the pendulum slowly moving to and fro, crossing the vertical wire at every vibration; on one occasion, just as it passes the wire, I touch the stud and start the watch. I allow the pendulum to make 300 vibrations, and as the silk arrives at the vertical wire for the 300th time, I promptly stop the watch; on reference I find that 241·6 seconds have elapsed since the time the watch was started. To avoid error, I repeat this experiment, with precisely the same result: 241·6 seconds are again required for the completion of 300 vibrations.

614. It is desirable to reckon the vibrations from the instant when the pendulum is at the middle of its stroke, rather than when it arrives at the end of the swing. In the former case the pendulum is moving with the greatest rapidity, and therefore the time of coincidence between the thread and the vertical wire can be observed with especial definiteness.

615. The time of a single vibration is found, by dividing 241·6 by 300, to be 0·805 second. This is certainly

correct to within a thousandth part of a second. We conclude that a pendulum whose length is $25''\cdot37 = 2\cdot114$, vibrates in 0·805 second; and from this we find that gravity at Dublin is $2'\cdot114 \times \left(\frac{3\cdot1416}{0\cdot805}\right)^2 = 32\cdot196$. This result agrees with one which has been determined by measurement made with every precaution.

Another method of measuring gravity by the pendulum will be described in the next lecture (Art. 637).

THE CYCLOID.

616. If the amplitude of the vibration of a circular pendulum bear a large proportion to the radius, the time of oscillation is slightly greater than if the amplitude be very small. The isochronism of the pendulum is only true for small arcs.

617. But there is a curve in which a weight may be made to move where the time of vibration is precisely the same, whatever be the amplitude. This curve is called a cycloid. It is the path described by a nail in the circumference of a wheel, as the wheel rolls along the ground. Thus, if a circle be rolled underneath the line A B (Fig. 85), a point on its circumference describes the cycloid A D C P B. The lower part of this curve does not differ very much from a circle whose centre is a certain point O above the curve.

618. Suppose we had a piece of wire carefully shaped to the cycloidal curve A D C P B, and that a ring could slide along it *without friction*, it would be found that, whether the ring be allowed to drop from C, P or B, it would fall to D precisely in the same time, and would then run up the wire to a distance from D on the other side

equal to that from which it had originally started. In the oscillations on the cycloid, the amplitude is absolutely without effect upon the time.

619. As a frictionless wire is impossible, we cannot adopt this method, but we can nevertheless construct a cycloidal pendulum in another way, by utilizing a property of the curve. O A (Fig. 85) as a half cycloid; in fact, O A is just the same curve as B D, but placed in a different position, so also is O B. If a string of length O D be suspended from the point O, and have a weight attached to it, the weight will describe the cycloid, provided that

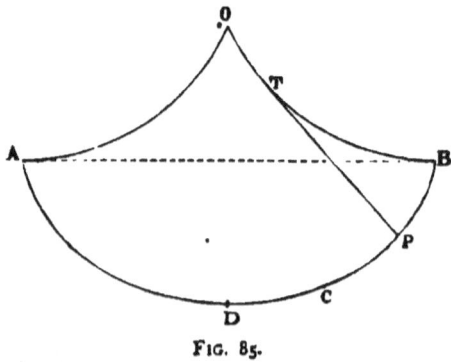

FIG. 85.

the string wrap itself along the arcs O A and O B; thus when the weight has moved from D to P, the string is wrapped along the curve through the space O T, the part T P only being free. This arrangement will always force the point P to move in the cycloidal arc.

620. We are now in a condition to ascertain experimentally, whether the time of oscillation in the cycloid is independent of the amplitude. We use for this purpose the apparatus shown in Fig. 86. D C E is the arc of the cycloid; two strings are attached at O, and equal weights A, B are suspended from them; C is the lowest point of

XVIII.] THE CYCLOID. 297

the curve. The time A will take to fall through the arc A C is of course half the time of its oscillation. If, therefore, I can show that A and B both take the same time to fall

Fig. 86.

down to C, I shall have proved that the vibrations are isochronous.

621. Holding, as shown in the figure, A in one hand and B in the other, I release them simultaneously, and you see the result,—they both meet at C: even if I bring A up to E,

and bring B down close to C, the result is the same. The motion of A is so rapid that it arrives at C just at the same instant as B. When I bring the two balls on the same side of C, and release them simultaneously, A overtakes B just at the moment when it is passing C. Hence, under all circumstances, the times of descent are equal.

622. It will be noticed that the string attached to the ball B, in the position shown in the figure, is almost as free as if it were merely suspended from O, for it is only when the ball is some distance from the lowest point that the side arcs produce any appreciable effect in curving the string. The ball swings from B to C nearly in a circle of which the centre is at O. Hence, in the circular pendulum, the vibrations when small are isochronous, for in that case the cycloid and the circle become indistinguishable.

LECTURE XIX.

THE COMPOUND PENDULUM AND THE COMPOSITION OF VIBRATIONS.

The Compound Pendulum.—The Centre of Oscillation.—The Centre of Percussion.—The Conical Pendulum.—The Composition of Vibrations.

THE COMPOUND PENDULUM.

623. PENDULOUS motion must now be studied in other forms besides that of the simple pendulum, which consists of a weight and a cord. Any body rotating about an axis may be made to oscillate by gravity. A body thus vibrating is called a *compound* pendulum. The ideal form, which consists of an indefinitely small weight attached to a perfectly flexible and imponderable string, is an abstraction which can only be approximately imitated in nature. It follows that every pendulum used in our experiments is strictly speaking compound.

624. The first pendulum of this class which we shall notice is that used in the common clock (Fig. 87). This consists of a wooden or steel rod A E, to which a brass or leaden bob B is attached. This pendulum is suspended by means of a steel spring C A, which being very flexible, allows

the vibration to be performed with considerable freedom. The use of the screw at the end will be explained in Art. 664. A pendulum like this vibrates isochronously, when the amplitude is small, but it is not easy to see precisely what is the length of the simple pendulum which would oscillate in the same time. In the first place, we are uncertain as to the virtual position of the point of suspension, for the spring, though flexible, will not yield at the point c to the same extent as a string; thus the effective point of suspension must be somewhat lower than c. The other extremity is still more uncertain, for the weight, so far from being a single point, is not exclusively in the neighbourhood of the bob, inasmuch as the rod of the pendulum has a mass that is appreciable. This form of pendulum cannot therefore be used where it is necessary to determine the length with accuracy.

625. When the length of a pendulum is to be measured, we must adopt other means of supporting it than that of suspension by a spring, as otherwise we cannot have a definite point from which to measure. To illustrate the mode that is to be adopted, I take here an iron bar 6' long and 1" square, which weighs 19 lbs. I wish to support this at one end so that it can vibrate freely, and at the same time have a definite point of suspension. I have here two small prisms of steel E (Fig. 88) fastened to a brass frame; the faces of the prisms meet at about an angle of 60° and form the edges about which the oscillation takes place: this frame and the edges can be placed on the end of the bar, and can be fixed there by tightening two nuts. The object of having the

FIG. 87.

edges on a sliding frame is that they may be applicable to different parts of the bar with facility. In some instruments used in experiments requiring extreme delicacy, the edges which are attached to the pendulum are supported upon plates of agate; they are to be adjusted on the same horizontal line, and the pendulum really vibrates about this line, as about an axis. For our purpose it will be sufficient to support the edges upon small pieces of steel. A B, Fig. 88, represents one side of the top of the iron bar; E is the edge projecting from it, with its edge perpendicular to the bar. C D is a steel plate bearing a knife edge on its upper surface; this piece of steel is firmly secured to the framework. There is of course a similar piece on the other side, supporting the other edge. The bar, thus delicately poised, will, when once started, vibrate backwards and forwards for an hour, as there is very little friction between the edges and the pieces which support them.

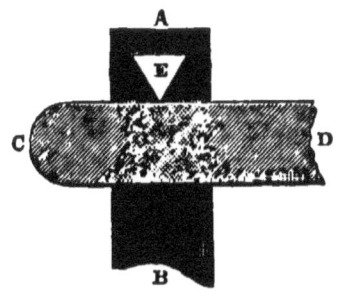

FIG. 88.

626. The general appearance of the apparatus, when mounted, is shown in Fig. 89. A B is the bar: at A the two edges are shown, and also the pieces of steel which support them. The whole is carried by a horizontal beam bolted to two uprights; and a glance at the figure will explain the arrangements made to secure the steadiness of the apparatus; the second pair of edges shown at B will be referred to presently (Art. 635).

627. This bar, as you see, vibrates to and fro; and we shall determine the length of a simple pendulum which would vibrate in the same period of time. The length might be

deduced by finding the time of vibration, and then calculating from Art. 606. This would be the most accurate mode of proceeding, but I have preferred to adopt a direct method which does not require calculation. A simple

Fig. 89.

pendulum, consisting of a fine cord and a small iron sphere C, is mounted behind the edge, Fig. 89. The point from which the cord is suspended lies exactly in the line of the two edges, and there is an adjustment for lengthening or shortening the cord at pleasure.

628. We first try with 6' of cord, so that the simple pendulum shall have the same length as the bar. Taking the ball in one hand and the bar in the other, I draw them aside, and you see, when they are released, that the bar performs two vibrations and returns to my hand before the ball. Hence the length of the isochronous simple pendulum is certainly less than the length of the bar; for a pendulum of that length is too slow.

629. I now shorten the cord until it is only half the length of the bar; and, repeating the experiment, we find that the ball returns before the bar, and therefore the simple pendulum is too short. Hence we learn that the isochronous pendulum is greater than half the length of the bar, and less than the whole length.

630. Let us finally try a simple pendulum two-thirds of the length of the bar. I make the experiment, and find that the ball and the bar return to my hand precisely at the same instant. Therefore two-thirds of the length of the bar is the length of the isochronous simple pendulum.

We may state generally that *the time of vibration of a uniform bar about one end equals that of a simple pendulum whose length is two-thirds of the bar;* no doubt the bar we have used is not strictly uniform, because of the edges; but in the positions they occupy, their influence on the time of vibrations is imperceptible.

632. For this rule to be verified, it is essentially necessary that the edges be properly situated on the bar; to illustrate this we may examine the oscillations of the small rod, shown at D (Fig. 89). This rod is also of iron $24'' \times 0''\cdot 5 \times 0''\cdot 5$, and it is suspended from a point near the centre by a pair of edges; if the edges could be placed so that the centre of gravity of the whole lay in the line of the edges, it is evident that the bar would rest indifferently

however it were placed, and would not oscillate. If then the edges be very near the centre of gravity, we can easily understand that the oscillations may be very slow, and this is actually the case in the bar D. By the aid of the stopwatch, I find that one hundred vibrations are performed in 248 seconds, and that therefore each vibration occupies 2·48 seconds. The length of the simple pendulum which has 2·48 seconds for its period of oscillation, is about 20′. Had the edges been at one end, the length of the simple pendulum would have been

$$24'' \times \tfrac{2}{3} = 16''.$$

A bar 72″ long will vibrate in a shorter time when the edge is 15″·2 from one end than when it has any other position. The length of the corresponding simple pendulum is 41″·6.

THE CENTRE OF OSCILLATION.

633. It appears that corresponding to each compound pendulum we have a specific length equal to that of the isochronous simple pendulum. To take as example the 6′ bar already described (Art. 625), this length is 4′. If I measure off from the edges a distance of 4′, and mark this point upon the bar, the point is called the *centre of oscillation*. More generally the centre of oscillation is found by drawing a line equal to the isochronous simple pendulum from the centre of oscillation through the centre of gravity.

634. In the bar D the centre of oscillation would be at a distance of 20′ below the edges; and in general the position will vary with the position of the edges.

635. In the 6′ bar B is the centre of oscillation. I take another pair of edges and place them on the bar, so that the line of the edges passes through B. I now lift the bar

XIX.] THE CENTRE OF OSCILLATION. 305

carefully and turn it upside down, so that the edges B rest upon the steel plates. In this position one-third of the bar is above the axis of suspension, and the remaining two-thirds below. A is of course now at the bottom of the bar, and is on a level with the ball, c: the pendulum is made to oscillate about the edges B, and the time of its vibration may be approximately determined by direct comparison with C, as already explained. I find that, when I allow c and the bar to swing together, they both vibrate precisely in the same time. You will remember, that when the ball was suspended by a string 4' long, its vibrations were isochronous with those of the bar when suspended from the edges A. Without having altered c, but having made the bar to vibrate about B, I find that the time of oscillation of the bar is still equal to that of C. Therefore, the period of oscillation about A is equal to that about B. Hence, when the bar is vibrating about B, its centre of oscillation must be 4' from B, that is, it must be at A: so that when the bar is suspended from A, the centre of oscillation is B; while, when the bar is suspended from B, the centre of oscillation is A. This is an interesting dynamical theorem. It may be more concisely expressed by saying that *the centre of oscillation and the centre of suspension are reciprocal.*

636. Though the proof that we have given of this curious law applies only to a uniform bar, yet the law is itself true in general, whatever be the nature of the compound pendulum.

637. We alluded in the last lecture (Art. 610) to the difficulty of measuring with accuracy the length of a simple pendulum; but the reciprocity of the centres of oscillation and suspension, suggested to the ingenious Captain Kater a method by which this difficulty could be

x

evaded. We shall explain the principle. Let one pair of edges be at A. Let the other pair of edges, B, be moved as near as possible to the centre of oscillation. We can test whether B has been placed correctly: for the time taken by the pendulum to perform 100 vibrations about A should be equal to the time taken to perform 100 vibrations about B. If the times are not quite equal, B must be moved slightly until the times are properly brought to equality. The length of the isochronous simple pendulum is then equal to the distance between the edges A and B; and this distance, from one edge to the other edge, presents none of the difficulties in its exact measurement which we had before to contend with: it can be found with precision. Hence, knowing the length of the pendulum and its time of oscillation, gravity can be found in the manner already explained (Art. 608).

638. I have adjusted the two edges of the 6' bar as nearly as I could at the centres of oscillation and suspension, and we shall proceed to test the correctness of the positions. Mounting the bar first by the edges at A, I set it vibrating. I take the stop-watch already referred to (Art. 612), and record the position of its hands. I then place my finger on the stud, and, just at the moment when the bar is at the middle of one of its vibrations, I start the watch. I count a hundred vibrations; and when the pendulum is again at the middle of its stroke, I stop the watch, and find it records an interval of 110·4 seconds. Thus the time of one vibration is 1·104 seconds. Reversing the bar, so that it vibrates about its centre of oscillation B, I now find that 110·0 is the time occupied by one hundred vibrations counted in the same manner as before; hence 1·100 seconds is the time of one vibration about B: thus, the periods of the vibrations are very nearly equal, as they differ only by $\frac{1}{250}$th part of a second.

639. It would be difficult to render the times of oscillation exactly equal by merely altering the position of B. In Kater's pendulum the two knife-edges are first placed so that the periods are as nearly equal as possible. The final adjustments are given by moving a small sliding-piece on the bar until it is found that the times of vibration about the two edges are identical. We shall not, however, use this refinement in a lecture experiment; I shall adopt the mean value of 1·102 seconds. The distance of the knife-edges is about 3″·992; hence gravity may be found from the expression (Art. 608)

$$3''·992 \times \left(\frac{3·1416}{1·102}\right)^2$$

The value thus deduced is 32′·4, which is within a small fraction of the true value.

640 With suitable precautions Kater's pendulum can be made to give a very accurate result. It is to be adjusted so that there shall be no perceptible difference in the number of vibrations in twenty-four hours, whichever edge be the axis of suspension: the distance between the edges is then to be measured with the last degree of precision by comparison with a proper standard.

THE CENTRE OF PERCUSSION.

641. The centre of oscillation in a body free to rotate about a fixed axis is identical with another remarkable point, called the *centre of percussion*. We proceed to examine some of the properties of a body thus suspended with reference to the effects of a blow. For the purposes of these experiments the method of suspension by edges is however quite unsuited.

642. We shall first use a rod suspended from a pin about

which the rod can rotate. A B, Fig. 90, is a pine rod 48" × 1" × 1", free to turn round B. Suppose this rod be hanging vertically at rest. I take a stick in my hand, and, giving the rod a blow, an impulsive shock will instantly be communicated to the pin at B; but the actual effect upon B will be very different according to the position at which the blow is given. If I strike the upper part of the rod at D, the action of A B upon the pin is a pressure to the left. If I strike the lower part at A, the pressure is to the right. But if I strike the point C, which is distant from B by two-thirds of the length of the rod, there is no pressure upon the pin. Concisely, for a blow below C, the pressure is to the right; for one above C, it is to the left; for one at C it is nothing.

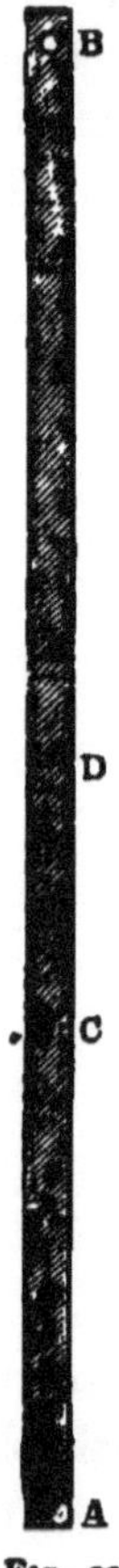

643. We can easily verify this by holding one extremity of a rod between the finger and thumb of the left hand, and striking it in different places with a stick held in the right hand; the pressure of the rod, when struck, will be so felt that the circumstances already stated can be verified.

644. A more visible way of investigating the subject is shown in Fig. 91. F B is a rod of wood, suspended from a beam by the string F G. A piece of paper is fastened to the rod at F by means of a small slip of wood clamped firmly to the rod; the other ends of this piece of paper are similarly clamped at P and Q.

FIG. 90.

645. When the rod receives a blow on the right-hand side at A, we find that the piece of paper is broken across at E, because the end F has been driven by the blow towards Q, and consequently caused the fracture of the paper at a place, E, where it had been specially narrowed.

XIX.] THE CENTRE OF PERCUSSION. 309

I remove the pieces of paper, and replace them by a new piece precisely similar. I now strike the rod at B,—a smart tap is all that is necessary,—and the piece of paper breaks at D. Finally replacing the pieces of paper by a third piece, I find that when I give the rod a tap (not a violent blow) at C, neither D nor E are broken.

646. This point C, where the rod can receive a blow without producing a shock upon the axis of suspension is the *centre of percussion*. We see, from its being two-thirds of the length of the rod distant from F, that it is identical with the centre of oscillation of the rod, if vibrating about knife-edges at F. It is true in general, whatever be the shape of the body, that the centre of oscillation is identical with the centre of percussion.

647. The principle embodied in the property of the centre of percussion has many practical applications. Every cricketer well knows that there is a part of his bat from which the ball flies without giving his hands

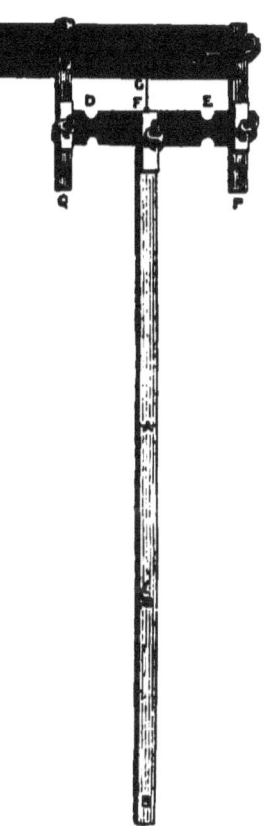

FIG. 91.

any unpleasant feeling. The explanation is simple. The bat is a body suspended from the hands of the batman; and if the ball be struck with the centre of percussion of the bat, there is no shock experienced. The centre of percussion in a hammer lies in its head, consequently a nail can receive a violent blow with perfect comfort to the hand which holds the handle.

THE CONICAL PENDULUM.

648. I have here a tripod (Fig. 92) which supports a heavy ball of cast iron by a string 6′ long. If I withdraw the ball from its position of rest, and merely release it, the ball vibrates to and fro, the string continues in the same plane, and the motion is that already discussed in the circular pendulum. If at the same instant that I release the ball, I impart to it a slight push in a direction not passing through the position of rest, the ball describes a curved path, returning to the point from which it started. This motion is that of the conical pendulum, because the string supporting the ball describes a cone.

649. In order to examine the nature of the motion, we can make the ball depict its own path. At the opposite point of the ball to that from which it is suspended, a hole is drilled, and in this I have fitted a camel's hair paintbrush filled with ink. I bring a sheet of paper on a drawing-board under the vibrating ball; and you see the brush traces an ellipse upon the paper, which I quickly withdraw.

650. By starting the ball in different ways, I can make it describe very different ellipses: here is one that is extremely long and narrow, and here another almost circular. When the magnitude of the initial velocity is properly adjusted, and its direction is perpendicular to the radius, I can make the string describe a right cone, and the ball a horizontal circle, but it requires some care and several trials in order to succeed in this. The ellipse may also become very narrow, so that we pass by insensible

XIX.   THE CONICAL PENDULUM.   311

gradations to the circular pendulum, in which the brush traces a straight line.

651. When the ball is moving in a circle, its velocity

FIG 92.

is uniform; when moving in an ellipse, its velocity is greatest at the extremities of the least axes of this ellipse, and least at the extremities of the greatest axes; but, when the ball is vibrating to and fro, as in the ordinary circular pendulum, the velocity is greatest at the middle of each vibration, and vanishes of course each time the pendulum attains the extremity of its swing. It is worthy of notice that under all circumstances the brush traces an ellipse upon the paper; for the circle and the straight line are only extreme cases, the one being a very round ellipse and the other a very thin one. If, however, the arc of vibration be large the movement is by no means so simple.

652. How are we to account for the elliptic movement? To do so fully would require more calculation than can be admitted here, but we may give a general account of the phenomenon.

Let us suppose that the ellipse A C B D, Fig. 93, is the path described by a particle when suspended by a string from a point vertically above o, the centre of the ellipse. To produce this motion I withdraw the particle from its position of rest at o to A. If merely released, the particle would swing over to B, and back again to A; but I do not simply release it, I impart a velocity impelling it in the direction A T. Through o draw C D parallel to A T. If I had taken the particle at o, and, without withdrawing it from its position of rest, had started it off in the direction o D, the particle would continue for ever to vibrate backwards and forwards from C to D. Hence, when I release the particle at A, and give it a velocity in the direction A T, the particle commences to move under the action of two distinct vibrations, one parallel to A B, the other parallel to C D, and we have to find the effect of these two vibrations impressed simultaneously upon the same particle. They are performed in the same time, since all vibrations are isochronous. We must conceive one motion starting from A towards o at the same moment that the other commences to start from o towards D. After the lapse of a short time, the body

XIX.] THE CONICAL PENDULUM. 313

has moved through A Y in its oscillation towards O, and in the same time through O Z in its oscillation towards D; it is therefore found at X. Now, when the particle has moved through a distance equal and parallel to A O, it must be found at the point D, because the motion from O to D takes the same time as from A to O. Similarly the body must pass through B, because the time occupied by going from A to B, would have been sufficient for the journey from O to D, and back again. The particle is found at P, because, after the

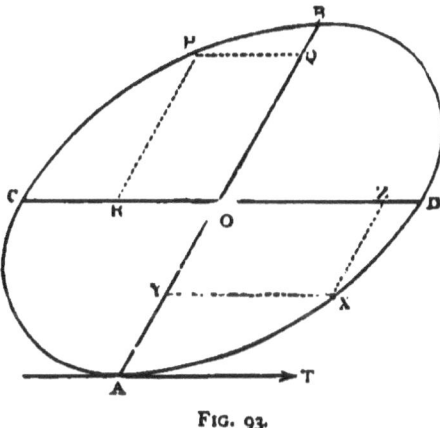

FIG. 93.

vibration returning from B has arrived at Q, the movement from D to O has travelled on to R. In this way the particle may be traced completely round its path by the composition of the two motions. It can be proved that for small motions the path is an ellipse, by reasoning founded upon the fact that the vibrations are isochronous.

653. Close examination reveals a very interesting circumstance connected with this experiment. It may be observed that the ellipse described by the body is not quite fixed in position, but that it gradually moves round in its plane.

Thus, in Fig. 92, the ellipse which is being traced out by the brush will gradually change its position to the dotted line shown on the board. The axis of the ellipse revolves in the same direction as that in which the ball is moving. This phenomenon is more marked with an ellipse whose dimensions are considerable in proportion to the length of the string. In fact, if the ellipse be very small, the change of position is imperceptible. The cause of this change is to be found in the fact already mentioned (Art. 598), that though the vibrations of a pendulum are very nearly isochronous, yet they are not absolutely so; the vibrations through a long arc taking a minute portion of time longer than those through a short arc.

This difference only becomes appreciable when the larger arc is of considerable magnitude with reference to the length of the pendulum.

654. How this causes displacement of the ellipse may be explained by Fig. 94. The particle is describing the figure A D C B in the direction shown by the arrows. This motion may be conceived to be compounded of vibrations A C and B D, if we imagine the particle to have been started from A with the right velocity perpendicular to O A. At the point A, the entire motion is for the instant perpendicular to O A; in fact, the motion is then exclusively due to the vibration B D, and there is no movement parallel to O A. We may define the extremity of the major axis of the ellipse to be the position of the particle, when the motion parallel to that axis vanishes. Of course this applies equally to the other extremity of the axis C, and similarly at the points B or D there is no motion of the particle parallel to B D.

655. Let us follow the particle, starting from A until it returns there again. The movement is compounded of

XIX.] THE COMPOSITION OF VIBRATIONS. 315

two vibrations, one from A to C and back again, the other along B D; from O to D, then from D to B, then from B to O, taking exactly double the time of one vibration from D to B. If the time of vibration along A C were exactly equal to that along B D, these two vibrations would bring the particle back to A precisely under the original circumstances. But they do not take place in the same time; the motion along A C takes a shade longer, so that, when the motion parallel to A C has ceased, the motion along D B has gone past O to a point Q, very near O. Let A P = O Q, and when the motion parallel to A C has vanished, the particle will be found at P; hence P must be the extremity of the major axis of the ellipse. In the next revolution, the extremity of the axis will advance a little more, and thus the ellipse moves round gradually.

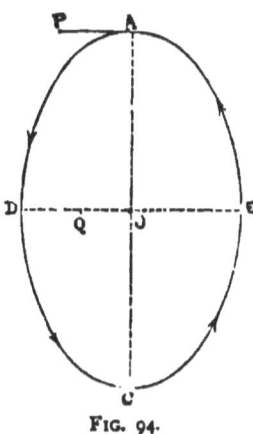

FIG. 94.

THE COMPOSITION OF VIBRATIONS.

656.—We have learned to regard the elliptic motion in the conical pendulum as compounded of two vibrations. The importance of the composition of vibrations justifies us in examining this subject experimentally in another way. The apparatus which we shall employ is represented in Fig. 95.

A is a ball of cast iron weighing 25 lbs., suspended from the tripod by a cord: this ball itself forms the support of another pendulum, B. The second pendulum is very light, being merely a globe of glass filled with sand. Through a

316   EXPERIMENTAL MECHANICS.   [LECT.

hole at the bottom of the glass the sand runs out upon a drawing-board placed underneath to receive it.

Thus the little stream of sand depicts its own journey upon

FIG. 96.

the drawing-board, and the curves traced out thus indicate the path in which the bob of the second pendulum has moved.

XIX.] THE COMPOSITION OF VIBRATIONS. 317

657. If the lengths of the two pendulums be equal, and their vibrations be in different planes, the curve described is an ellipse, passing at one extreme into a circle, and at the other into a straight line. This is what we might have expected, for the two vibrations are each performed in the same time, and therefore the case is analogous to that of the conical pendulum of Art. 648.

658. But the curve is of a very different character when the cords are unequal. Let us study in particular the case in which the second pendulum is only one-fourth the length of the cord supporting the iron ball. This is the experiment represented in Fig. 95. The form of the path delineated by the sand is shown in Fig. 96. The arrow-heads placed upon the curve show the direction in which it is traced. Let us suppose that the formation of the figure commences at A; it then goes on to B, to O, to C, to D, and back to A: this shows us that the bob of the lower pendulum must have performed two vibrations up and down, while that of the upper has made one right and left. The motion is thus compounded of two vibrations at right angles, and the time of one is half that of the other.

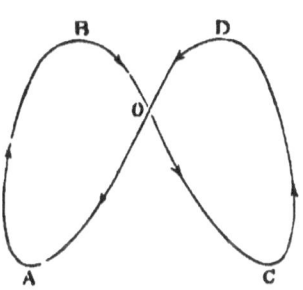

Fig. 96.

The time of vibration is proportional to the square root of the length; and, since the lower pendulum is one-fourth the length of the upper, its time of vibration is one-half that of the upper. In this experiment, therefore, we have a confirmation of the law of Art. 605.

LECTURE XX.

THE MECHANICAL PRINCIPLES OF A CLOCK.

Introduction.—The Compensating Pendulum.—The Escapement.—The Train of Wheels.—The Hands.—The Striking Parts.

INTRODUCTION.

659. WE come now to the most important practical application of the pendulum. The vibrations being always isochronous, it follows that, if we count the number of vibrations in a certain time, we shall ascertain the duration of that time. It is simply the product of the number of vibrations with the period of a single one. Let us take a pendulum 39·139 inches long; which will vibrate exactly once a second in London, and is therefore called a seconds pendulum (See Art. 607). If I set one of these pendulums vibrating, and contrive mechanism by which the number of its vibrations shall be recorded, I have a means of measuring time. This is of course the principle of the common clock: the pendulum vibrates once a second and the number of vibrations made from one epoch to another epoch is shown by the hands of the clock. For

LECT. XX.] THE COMPENSATING PENDULUM. 319

example, when the clock tells me that 15 minutes have elapsed, what it really shows is that the pendulum has made 60 × 15 = 900 vibrations, each of which has occupied one second.

660. One duty of the clock is therefore to count and record the number of vibrations, but the wheels and works have another part to discharge, and that is to sustain the motion of the pendulum. The friction of the air and the resistance experienced at the point of suspension are forces tending to bring the pendulum to rest; and to counteract the effect of these forces, the machine must be continually invigorated with fresh energy. This supply is communicated by the works of the clock, which will be sufficiently described presently.

661. When the weight driving the clock is wound up, a store of energy is communicated which is doled out to the pendulum in a very small impulse, at every vibration. The clock-weight is just large enough to be able to counterbalance the retarding forces when the pendulum has a proper amplitude of vibration. In all machines there is some energy lost in maintaining the parts in motion in opposition to friction and other resistances; in clocks this represents the whole amount of the force, as there is no external work to be performed.

THE COMPENSATING PENDULUM.

662. The actual length of the pendulum used, depends upon the purposes for which the clock is intended, but it is essential for correct performance that the pendulum should vibrate at a constant rate; a small irregularity in this respect may appreciably affect the indications of the clock. If the pendulum vibrates in 1·001 seconds instead of in one second, the clock loses one

thousandth of a second at each beat; and, since there are 86,400 seconds in a day, it follows that the pendulum will make only 86,400 − 86·3 vibrations in a day, and therefore the clock will lose 86·3 seconds, or nearly a minute and a half daily.

663. For accurate time-keeping it is therefore essential that the time of vibration shall remain constant. Now the time of vibration depends upon the length, and therefore it is necessary that the length of the pendulum be absolutely unalterable. If the length of the pendulum be changed even by one-tenth of an inch, the clock will lose or gain nearly two minutes daily, according to whether the pendulum has been made longer or shorter. In general we may say that, if the alteration in the length amount to k thousandths of an inch, the number of seconds gained or lost per day is $1\cdot 103 \times k$ with a seconds pendulum.

664. This explains the practice of raising the bob of the pendulum when the clock is going too slow or lowering it when going too fast. If the thread of the screw used in doing this have twenty threads to the inch; then one complete revolution of the screw will raise the bob through 50 thousandths of an inch, and therefore the effect on the rate will be $1\cdot 103 \times 50 = 55$ nearly. Thus, the rate of the clock will be altered by about 55 seconds daily. Whatever be the screw, its effect can be calculated by the simple rule expressed as follows. Divide 1103 by the number of threads to the inch; the quotient is the number of seconds that the clock can be made to gain or lose daily by one revolution of the screw on the bob of the pendulum.

665. Let us suppose that the length of the pendulum has been properly adjusted so that the clock keeps accurate time. It is necessary that the pendulum should not alter in length. But there is an ever-present cause tending to

xx.]   THE COMPENSATING PENDULUM.   321

change it. That cause is the variation of temperature. We shall first illustrate by actual experiment the well known law that bodies expand under the action of heat; then we shall consider the irregularities thus introduced into the motion of the pendulum; and, finally, we shall point out means by which these irregularities may be effectually counteracted.

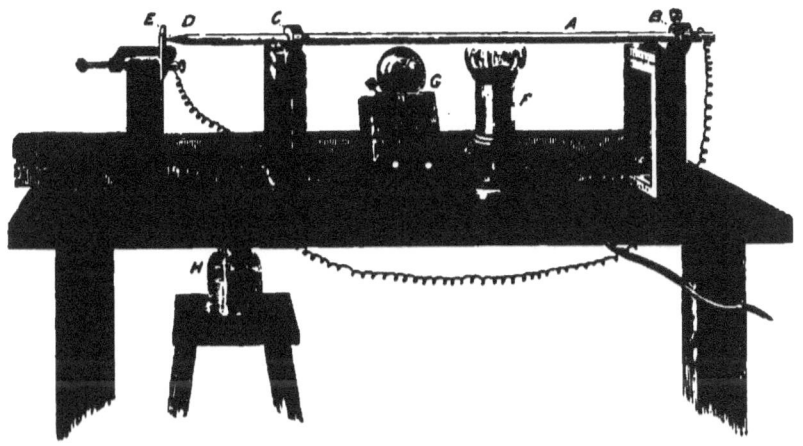

FIG. 97.

666. We have here a brass bar a yard long; it is at present at the temperature of the room. If we heat the bar over a lamp, it becomes longer; but upon cooling, it returns to its original dimensions. These alterations of length are very small, indeed too small to be perceived except by careful measurement; but we shall be able to demonstrate in a simple way that elongation is the consequence of increased temperature. I place the bar A D in the supports shown in Fig. 97. It is firmly secured at B by means of a binding screw, and passes quite freely through C; if the bar

Y

elongate when it is heated by the lamp, the point D must approach nearer to E. At H is an electric battery, and G is a bell rung by an electric current. One wire of the battery connects H and G, another connects G with E, and a third connects H with the end of the brass rod B. Until the electric current becomes completed, the bell remains dumb, the current is not closed until the point touches E: when this is the case, the current rushes from the battery along the bar, then from D to E, from that through the bell, and so back to the battery. At present the point is not touching E, though extremely close thereto. Indeed if I press E towards the point, you hear the bell, showing that the circuit is complete; removing my finger, the bell again becomes silent, because E springs back, and the current is interrupted.

667. I place the lamp under the bar: which begins to heat and to elongate; and as it is firmly held at B, the point gradually approaches E: it has now touched E; the circuit is complete, and the bell rings. If I withdraw the lamp, the bar cools. I can accelerate the cooling by touching the bar with a damp sponge; the bar contracts, breaks the circuit, and the bell stops: heating the bar again with the lamp, the bell again rings, to be again stopped by an application of the sponge. Though you have not been able to see the process, your ears have informed you that heat must have elongated the bar, and that cold has produced contraction.

668. What we have proved with respect to a bar of brass, is true for a bar of any material; and thus, whatever be the substance of which a pendulum is made, a simple uncompensated rod must be longer in hot weather than in cold weather: hence a clock will generally have a tendency to go faster in winter than in summer.

669. The amount of change thus produced is, it is true,

very small. For a pendulum with a steel rod, the difference of temperature between summer and winter would cause a variation in the rate of five seconds daily, or about half a minute in the week. The amount of error thus introduced is of no great consequence in clocks which are only intended for ordinary use; but in astronomical clocks, where seconds or even portions of a second are of importance, inaccuracies of this magnitude would be quite inadmissible.

670. There are, it is true, some substances—for example, ordinary timber—in which the rate of expansion is less than that of steel; consequently, the irregularities introduced by employing a pendulum with a wooden rod are less than those of the steel pendulum we have mentioned; but no substance is known which would not originate greater variations than are admissible in the performance of an astronomical clock.

We must, therefore, devise some means by which the effect of temperature on the length of a pendulum can be avoided. Various means have been proposed, and we shall describe one of the best and simplest.

671. The mercurial pendulum (Fig. 98) is frequently used in clocks intended to serve as standard timekeepers. The rod by which the pendulum is suspended is made of steel; and the bob consists of a glass jar of mercury. The distance of the centre of gravity of the mercury from the point of suspension may practically be considered as the length of the pendulum. The rate of expansion of mercury is about sixteen times that of steel: hence, if the bob be formed of a column of mercury one-eighth part of the length of the steel rod, the compensation would be complete. For, suppose the temperature of the pendulum be raised, the steel rod would be lengthened,

and therefore the vase of mercury would be lowered; on the other hand, the column of mercury would expand by an amount double that of the steel rod: thus the centre of the column of mercury would be elevated as much as the steel was elongated; hence the centre of the mercury is raised

FIG. 98.

by its own expansion as much as it is lowered by the expansion of the steel, and therefore the effective length of the pendulum remains unaltered. By this contrivance the time of oscillation of the pendulum is rendered independent of the temperature. The bob of the mercurial pendulum is shown in Fig. 98. The screw is for the purpose of raising or lowering the entire vessel of mercury in order to make the rate correct in the first instance.

THE ESCAPEMENT.

672. Practical skill as well as some theoretical investigation has been expended upon that part of a clock which is called the *escapement*, the excellence of which is essential to the correct performance of a timepiece. The pendulum must have its motion sustained by receiving an impulse at every vibration: at the same time it is desirable that the vibration should be hampered as little as possible by mechanical connection. The isochronism on which the time-keeping depends is in strictness only a characteristic of oscillations performed with a total freedom from constraint of every description; hence we must endeavour to approximate

XX.]   THE ESCAPEMENT.   325

the clock pendulum as nearly as possible to one which is swinging quite freely. To effect this, and at the same time to maintain the arc of vibration tolerably constant, is the property of a good escapement.

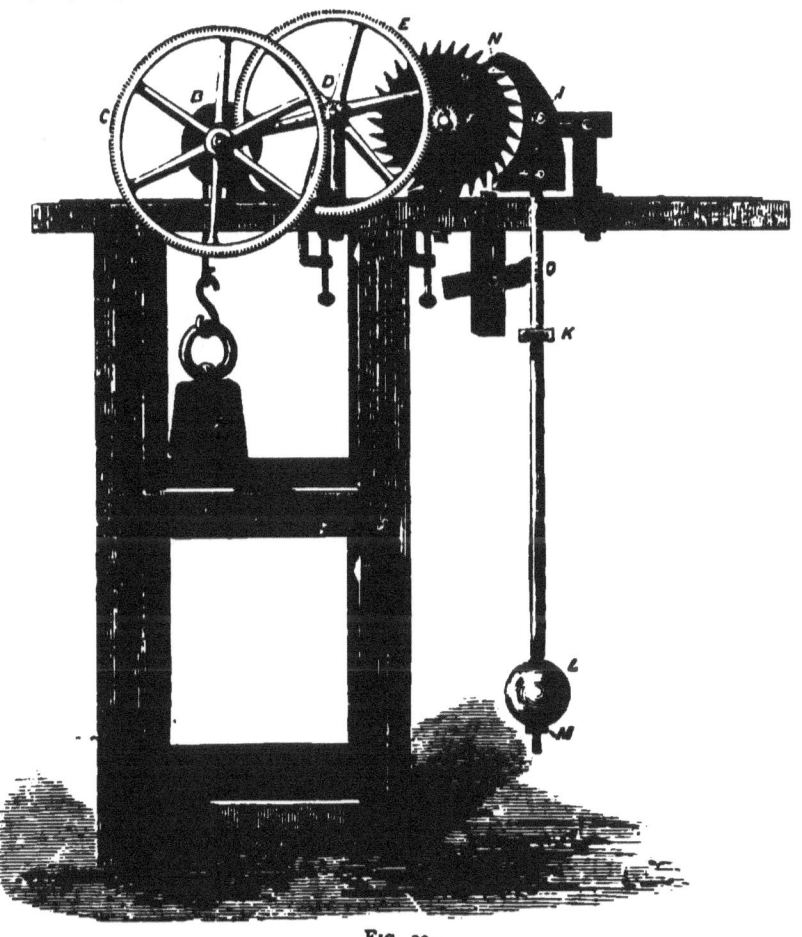

Fig. 99.

673. A common form of escapement is shown in Fig. 99. The arrangement is no doubt different from that actually found in a clock; but I have constructed the machine in

this way in order to show clearly the action of the different parts. G is called the escapement-wheel: it is surrounded by thirty teeth, and turns round once when the pendulum has performed sixty vibrations,—that is, once a minute. I represents the escapement; it vibrates about an axis and carries a fork at K which projects behind, and the rod of the pendulum hangs between its prongs. The pendulum is itself suspended from a point O. At N, H are a pair of polished surfaces called the pallets: these fulfil a very important function.

674. The escapement-wheel is constantly urged to turn round by the action of the weight and train of wheels, of which we shall speak presently; but the action of the pallets regulates the rate at which the wheel can revolve. When a tooth of the wheel falls upon the pallet N, the latter is gently pressed away: this pressure is transmitted by the fork to the pendulum; as N moves away from the wheel, the other pallet H approaches the wheel; and by the time N has receded so far that the tooth slips from it, H has advanced sufficiently far to catch the tooth which immediately drops upon H. In fact, the moment the tooth is free from N, the wheel begins to revolve in consequence of the driving weight; but it is quickly stopped by another tooth falling on H: and the noise of this collision is the well-known tick of the clock. The pendulum is still swinging to the left when the tooth falls on H. The pressure of the tooth then tends to push H outwards, but the inertia of the pendulum in forcing H inwards is at first sufficient to overcome the outward pressure arising from the wheel; the consequence is that, after the tooth has dropped, the escapement-wheel moves back a little, or "recoils," as it is called. If you look at any ordinary clock, which has a second-hand, you will notice

that after each second is completed the hand recoils before starting for the next second. The reason of this is, that the second-hand is turned directly by the escapement-wheel, and that the inertia of the pendulum causes the escapement-wheel to recoil. But the constant pressure of the tooth soon overcomes the inertia of the pendulum, and H is gradually pushed out until the tooth is able to "escape"; the moment it does so the wheel begins to turn round, but is quickly brought up by another tooth falling on N, which has moved sufficiently inwards.

The process we have just described then recurs over again. Each tooth escapes at each pallet, and the escapements take place once a second; hence the escapement-wheel with thirty teeth will turn round once in a minute.

675. When the tooth is pushing N, the pendulum is being urged to the left; the instant this tooth escapes, another tooth falls on H, and the pendulum, ere it has accomplished its swing to the left, has a force exerted upon it to bring it to the right. When this force and gravity combined have stopped the pendulum, and caused it to move to the right, the tooth soon escapes at H, and another tooth falls on N, then retarding the pendulum. Hence, except during the very minute portion of time that the wheel turns after one escapement, and before the next tick, the pendulum is never free; it is urged forwards when its velocity is great, but before it comes to the end of its vibration it is urged backwards; this escapement does not therefore possess the characteristics which we pointed out (Art. 672) as necessary for a really good instrument. But for ordinary purposes of time-keeping, the recoil escapement works sufficiently well, as the force which acts upon the pendulum is in reality extremely small. For the refined applications of the

astronomical clock, the performance of a recoil-escapement is inadequate.

The obvious defect in the recoil is that the pendulum is retarded during a portion of its vibration; the impulse forward is of course necessary, but the retarding force is useless and injurious.

676. The "dead-beat" escapement was devised by the celebrated clockmaker Graham, in order to avoid this difficulty. If you observe the second-hand of a clock, controlled by this escapement, you will understand why it is called the dead beat: there is no recoil; the second-hand moves quickly over each second, and remains there fixed until it starts for the next second.

The wheel and escapement by which this effect is produced is shown in Fig. 100. A and B are the pallets, by the action of the teeth on which the motion is given to the crutch, which turns about the centre O; from the axis through this centre the fork descends, so that as the crutch is made to vibrate to and fro by the wheel, the fork is also made to vibrate, and thus sustain the motion of the pendulum. The essential feature in which the dead-beat escapement differs from the recoil escapement is that when the tooth escapes from the pallet A, the wheel turns: but the tooth which in the recoil escapement would have fallen on the other pallet, now falls on a surface D, and not on the pallet B. D is part of a circle with its centre at O, the centre of motion; consequently, the tooth remains almost entirely inert so long as it remains on the circular arc D.

677. There is thus no recoil, and the pendulum is allowed to reach the extremity of its swing to the right unretarded; but when the pendulum is returning, the crutch moves until the tooth passes from the circular arc D on to the pallet B: instantly the tooth slides down the pallet, giving the crutch

THE ESCAPEMENT.

an impulse, and escaping when the point has traversed B. The next tooth that comes into action falls upon the circular arc C, of which the centre is also at O; this tooth likewise remains at rest until the pendulum has finished its swing, and has commenced its return; then the tooth slides down A, and the process recommences as before.

FIG. 100.

678. The operations are so timed that the pendulum receives its impulse (which takes place when a tooth slides down a pallet) precisely when the oscillation is at the point of greatest velocity; the pendulum is then unacted upon till it reaches a similar position in the next vibration. This impulse at the middle of the swing does not affect the time of vibration.

679. There is still a small frictional force acting to retard the pendulum. This arises from the pressure of the teeth upon the circular arcs, for there is a certain amount of friction, no matter how carefully the surfaces may be polished. It is not however found practically to be a source of appreciable irregularity.

In a clock furnished with a dead-beat escapement and a mercurial pendulum, we have a superb time-keeper.

THE TRAIN OF WHEELS.

680. We have next to consider the manner in which the supply of energy is communicated to the escapement-wheel, and also the mode in which the vibrations of the pendulum are counted. A train of wheels for this purpose is shown in Fig. 99. The same remark may be made about this train that we have already made about the escapement,—namely, that it is more designed to explain the principle clearly than to show the actual construction of a clock.

681. The weight A which animates the whole machine is attached to a rope, which is wound around a barrel B; the process of winding up the clock consists in raising this weight. On the same axle as the barrel B is a large tooth-wheel C; this wheel contains 200 teeth. The wheel C works into a pinion D, containing 20 teeth; consequently, when the wheel C has turned round once, the pinion D has turned round ten times. The large wheel E is on the same axle with the pinion D, and turns with D; the wheel E contains 180 teeth, and works into the pinion F, containing 30 teeth: consequently when E has gone round once, F will have turned round six times; and therefore, when the wheel C and the barrel B have made one revolution, the pinion F will have gone round sixty times; but the wheel G is on the same shaft as the pinion F, and therefore, for every

sixty revolutions of the escapement-wheel, the wheel C will have gone round once. We have already shown that the escapement-wheel goes round once a minute, and hence the wheel C must go round once in an hour. If therefore a hand be placed on the same axle with C, in front of a clock dial, the hand will go completely round once an hour; that is, it will be the minute-hand of the clock.

682. The train of wheels serves to transmit the power of the descending weight and thus supply energy to the pendulum. In the clock model you see before you, the weight sustaining the motion is 56 lbs. The diameter of the escapement-wheel is about double that of the barrel, and the wheel turns round sixty times as fast as the barrel; therefore for every inch the weight descends, the circumference of the escapement-wheel must move through 120 inches. From the principle of work it follows that the energy applied at one end of a machine equals that obtained from the other, friction being neglected. The force of 56 lbs. is therefore, reduced to the one hundred-and-twentieth part of its amount at the circumference of the escapement-wheel. And as the friction is considerable; the actual force with which each tooth acts upon the pallet is only a few ounces.

683. In a good clock an extremely minute force need only be supplied to the pendulum, so that, notwithstanding 86,400 vibrations have to be performed daily, one winding of the clock will supply sufficient energy to sustain the motion for a week

THE HANDS.

684. We shall explain by the model shown in Fig. 101, how the hour-hand and the minute-hand are made to revolve with different velocities about the same dial

EXPERIMENTAL MECHANICS. [LECT.

G is a handle by which I can turn round the shaft which carries the wheel F, and the hand B. There are 20 teeth in F, and it gears into another wheel, E, containing 80 teeth; the shaft which is turned by E carries a third wheel D, containing 25 teeth, and D works with a fourth C, containing 75 teeth. C is capable of turning freely round the shaft, so that the motion of the shaft does not affect it, except through the intervention of the wheels E, F, and D. To C another hand A is attached, which therefore turns round simultaneously with C. Let us compare the

FIG. 101.

motion of the two hands A and B. We suppose that the handle G is turned twelve times; then, of course, the hand B, since it is on the shaft, will turn twelve times. The wheel F also turns twelve times, but E has four times the number of teeth that A has, and therefore, when F has gone round four times, E will only have gone round once: hence, when F has revolved twelve times, E will have gone round three times. D turns with E, and therefore the twelve revolutions of the handle will have turned D round three times, but since C has 75 teeth and D 25 teeth, C will have only made one revolution, while D has made three revolutions; hence the hand A will have made only one revolution, while the hand B has made twelve revolutions.

We have already seen (Art. 681) how, by a train of wheels, one wheel can be made to revolve once in an hour. If that wheel be upon the shaft instead of the handle G, the hand B will be the minute-hand of the clock, and the hand A the hour-hand.

685. The adjustment of the numbers of teeth is important, and the choice of wheels which would answer is limited. For since the shafts are parallel, the distance between the centres of F and E must equal that between the centres of C and of D. But it is evident that the distance from the centre of F to the centre of E is equal to the sum of the radii of the wheels F and E. Hence the sum of the radii of the wheels F and E must be equal to the sum of the radii of C and D. But the circumferences of circles are proportional to their radii, and hence the sum of the circumferences of F and E must equal that of C and D; it follows that the sum of the teeth in E and F must be equal to the sum of the teeth in C and D. In the present case each of these sums is one hundred.

686. Other arrangements of wheels might have been devised, which would give the required motion; for example, if F were 20, as before, and E 240, and if C and D were each equal to 130, the sum of the teeth in each pair would be 260. E would only turn once for every twelve revolutions of F, and C and D would turn with the same velocity as E; hence the motion of the hand A would be one-twelfth that of B. This plan requires larger wheels than the train already proposed.

THE STRIKING PARTS.

687. We have examined the essential features of the

going parts of the clock; to complete our sketch of this instrument we shall describe the beautiful mechanism by which the striking is arranged. The model which we represent in Fig. 102 is, as usual, rather intended to illustrate the principles of the contrivance than to be an exact counterpart of the arrangement found in clocks. Some of the details are not reproduced in the model; but enough is shown to explain the principle, and to enable the model to work.

688. When the hour-hand reaches certain points on the dial, the striking is to commence; and a certain number of strokes must be delivered. The striking apparatus has both to initiate the striking and to control the number of strokes; the latter is by far the more difficult duty. Two contrivances are in common use; we shall describe that which is used in the best clocks.

689. An essential feature of the striking mechanism in the repeating clock is the snail, which is shown at B. This piece must revolve once in twelve hours, and is, therefore, attached to an axle which performs its revolution in exactly the same time as the hour-hand of the clock. In the model, the striking gear is shown detached from the going parts, but it is easy to imagine how the snail can receive this motion. The margin of the snail is marked with twelve steps, numbered from one to twelve. The portions of the margin between each pair of steps is a part of the circumference of a circle, of which the axis of the snail is the centre. The correct figuring of the snail is of the utmost importance to the correct performance of the clock. Above the snail is a portion of a toothed wheel, F, called the rack; this contains about fourteen or fifteen teeth. When this wheel is free, it falls down until a pin comes in contact with the snail at B.

690. The distance through which the rack falls depends upon the position of the snail; if the pin come in contact with the part marked I., as it does in the figure, the rack will descend but a small distance, while, if the pin fall on the part marked VII., the rack will have a longer fall: hence as the snail changes its position with the successive hours, so the distance through which the rack falls changes also. The snail is so contrived that at each hour the rack falls on a lower step than it does in the preceding hour; for example, during the hour of three o'clock, the rack would, if allowed to fall, always drop upon the part of the snail marked III., but, when four o'clock has arrived, the rack would fall on the part marked IV.; it is to insure that this shall happen correctly that such attention must be paid to the form of the snail.

691. A is a small piece called the "gathering pallet": it is so placed with reference to the rack that, at each revolution of A, the pallet raises the rack one tooth. Thus, after the rack has fallen, the gathering pallet gradually raises it.

692. On the same axle as the gathering pallet, and turning with it, is another piece C, the object of which is to arrest the motion when the rack has been raised sufficiently. On the rack is a projecting pin; the piece C passes free of this pin until the rack has been lifted to its original height, when C is caught by the pin, and the mechanism is stopped. The magnitude of the teeth in the rack is so arranged with reference to the snail, that the number of lifts which the pallet must make in raising the rack is equal to the number marked upon the step of the snail upon which the rack had fallen; hence the snail has the effect of controlling the number of revolutions which the gathering pallet can make. The rack is retained by a detent F, after being raised each tooth.

693. The gathering pallet is turned by a small pinion of 27 teeth, and the pinion is worked by the wheel C, of 180 teeth. This wheel carries a barrel, to which a movement of rotation is given by a weight, the arrangement of which is evident: a second pinion of 27 teeth on the same axle with D is also turned by the large wheel C. Since these pinions are equal, they revolve with equal velocities. Over D the bell I is placed; its hammer E is so arranged that a pin attached to D strikes the bell once in every revolution of D. The action will now be easily understood. When the hour-hand reaches the hour, a simple arrangement raises the detent F; the rack then drops; the moment the rack drops, the gathering pallet commences to revolve and raises up the rack; as each tooth is raised a stroke is given to the bell, and thus the bell strikes until the piece C is brought to rest against the pin.

694. The object of the fan H is to control the rapidity of the motion: when its blades are placed more or less obliquely, the velocity is lessened or increased.

APPENDIX I.

The formulæ in the tables on p. 73 and after can be deduced by two methods,—one that of graphical construction, the other that of least squares. The first method is the more simple and requires but little calculation; though neatness and care are necessary in constructing the diagrams. The second method will be described for the benefit of those who possess the requisite mathematical knowledge. The formulæ used in the preparation of the tables have been generally deduced from the method of least squares, as the results are to a slight, though insignificant, extent more accurate than those of the method of graphical construction. This remark will explain why the figures in some of the formulæ are carried to a greater number of places of decimals than could be obtained by the other method.

We shall confine the numerical examples to Tables III. and IV., and show how the formulæ of these tables have been deduced by the two different methods.

Tables V., XIV., XVI., XXI., are to be found in the same manner as Table III.; and Tables VI., IX., X., XI., XV., XVII., XVIII., XIX., XX., XXI., XXII., in the same manner as Table IV.

THE METHOD OF GRAPHICAL CONSTRUCTION.

TABLE III.

A horizontal line APS, shown on a diminished scale in Fig. 103, is to be neatly drawn upon a piece of cardboard about 14" × 6". A scale which reads to the hundreth of an inch is to be used in the construction of the figure. A pocket lens will be found convenient in reading the small divisions. By means of a pair of compasses and the scale, points are to be marked upon the line APS, at distances 1″·4, 2″·8, 4″·2, 5″·6, 7″·0, 8″·4, 9″·8, 11″·2 from the origin A. These distances correspond to the magnitudes of the loads placed upon the slide on the scale of 0″·1 to 1 lb. Perpendiculars to APS are to be erected at the points marked, and distances F_1, F_2, F_3, &c. set off upon these perpendiculars. These distances are to be equal, on the adopted scale, to the frictions for the corresponding loads. For example, we see from Table III., Experiment 3, that when the load upon the slide is 42 lbs., the friction is 12·2 lbs.; hence the point F_3 is found by measuring a distance 4″·2 from A, and erecting a

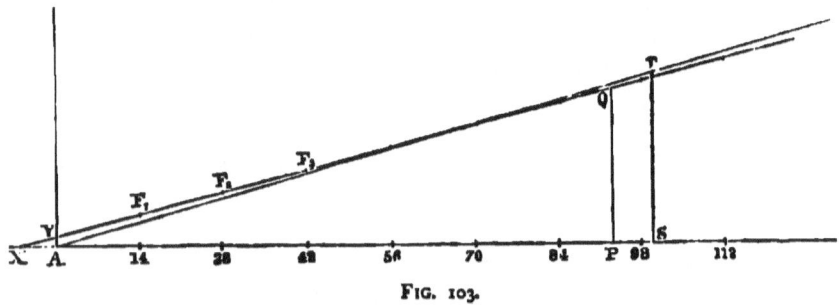

FIG. 103.

perpendicular 1″·22. Thus, for each of the loads a point is determined. The positions of these points should be indicated by making each of them the centre of a small circle 0″·1 diameter. These circles, besides neatly defining the points, will be useful in a subsequent part of the process.

It will be found that the points F_1, F_2, &c. are very nearly in a straight line. We assume that, if the apparatus and observations were perfect, the points would lie exactly in a straight line. The object of the construction is to determine the straight line, which on the whole is most close to all the points. If it be true that the friction is proportional to the pressure, this line

APPENDIX.

should pass through the origin A, for then the perpendicular which represents the friction is proportional to the line cut off from A, which represents the load. It will be found that a line AT can be drawn through the origin A, so that all the points are in the immediate vicinity of this line, if not actually upon it. A string of fine black silk about 15" long, stretched by a bow of wire or whalebone, is a convenient straight-edge for finding the required line. The circles described about the points F_1, F_2, &c. will facilitate the placing of the silk line as nearly as possible through all the points. It will not be found possible to draw a line through A, which shall intersect all the circles; the best line passes below but very near to the circles round F_1, F_2, F_3, F_4, touches the circle about F_5, intersects the circles about F_6 and F_7, and passes above the circle round F_8. The line should be so placed that its depth below the point which is most above it, is equal to the height at which it passes above the point which is most below it.

From A measure AS, a length of 10", and erect the perpendicular S T. We find by measurement that ST is 2"·7. If, then, we suppose that the friction for any load is really represented by the distance cut off by the line AT upon the perpendicular, it follows that

$$F : R : : 2'''7 : 10''.$$
$$\text{or } F = 0.27\, R.$$

This is the formula from which Table III. has been constructed.

Table IV.

By a careful application of the silk bow-string, X Y Q can be drawn, which, itself in close proximity to A, passes more nearly through F_1, F_2, &c. than is possible for any line which passes exactly through A. X Y Q will be found not only to intersect all the small circles, but to cut off a considerable arc from each. Measure off X P a distance of 10", and erect the perpendicular

P Q; then, if R be the load, and F the corresponding friction, we must have from similar triangles—

$$\frac{F - \frac{AY}{0''\cdot 1} \times 1 \text{ lb.}}{R} = \frac{PQ}{PX}.$$

By measurement it is found that $AY = 0''\cdot 14$, and $PQ = 2''\cdot 53$.

We have, therefore,

$$F = 1\cdot 4 + 0\cdot 253\, R.$$

This is practically the same formula as

$$F = 1\cdot 44 + 0\cdot 252\, R,$$

from which the table has been constructed. In fact, the column of calculated values of the friction might have been computed from the former, without appreciably differing from what is found in the table.

THE METHOD OF LEAST SQUARES.

Table III.

Let k be the coefficient of friction. It is impossible to find any value for k which will satisfy the equation,

$$F - kR = 0,$$

for all the observed pairs of values of F and R. We have then to find the value for k which, upon the whole, best represents the experiments. $F - kR$ is to be as near zero as possible for each pair of values of F and R.

In accordance with the principle of least squares, it is well known to mathematicians, the best value of k is that which makes

$$(F_1 - k R_1)^2 + (F_2 - k R_2)^2 + \&c. + (F_m - k R_m)^2$$

a minimum where F_1 and R_1, F_2 and R_2 &c. are the simultaneous values of F and R in the several experiments.

APPENDIX. 343

In fact, it is easy to see that, if this quantity be small, each of the essentially positive elements,

$$(F_1 - k R_1)^2, \&c.$$

of which it is composed, must be small also, and that therefore

$$F - k R$$

must always be nearly zero.

Differentiating the sum of squares and equating the differential coefficeint to zero, we have according to the usual notation,

$$\Sigma R_1 (F_1 - k R_1) = 0;$$
$$\text{whence } k = \frac{\Sigma R_1 F_1}{\Sigma R_1^2}.$$

The calculation of k becomes simplified when (as is generally the case in the tables) the loads $R_1, R_2, \&c., R_m$ are of the form,

$$N, 2 N, 3 N, \&c., m N.$$

In this case,

$$\Sigma R_1 F_1 = N (F_1 + {}_2F_2 + 3 F_3 + \&c. + m F_m).$$
$$\Sigma R_1^2 = N^2 (1^2 + 2^2 + \&c. + m^2)$$
$$= N^2 \frac{m (m + 1) (2 m + 1)}{6}$$
$$\therefore k = 6 \frac{(F + 2 F_2 + \&c. + m F_m)}{N m (m + 1) (2 m + 1)}.$$

In the case of Table III.

$$m = 8, N = 14,$$
$$F_1 + 2 F + 3 F_3 + m F_m = 770\cdot9;$$
$$\text{whence } k = 0\cdot27.$$

Thus the formula $F = 0\cdot27 R$ is deduced both by the method of least squares, and by the method of graphical construction.

TABLE IV.

The formula for this table is to be deduced from the following considerations.

No values exist for x and y, so that the equation

$$F = x + y R$$

shall be satisfied for all pairs of values of F and R, but the best values for x and y are those which make

$$(F_1 - x - y R_1)^2 + (F_2 - x - y R_2)^2 + \&c. + (F_m - x - y R_m)^2$$

a minimum.

Differentiating with respect to x and y, and equating the differential coefficients to zero, we have

$$\Sigma (F_1 - x - y R_1) = 0,$$
$$\Sigma R_1 (F_1 - x - y R_1) = 0.$$

This gives two equations for the determination of x and y. Suppose, as is usually the case, the loads be of the form,

$$N, 2N, 3N, 4N \&c.\ mN,$$

and making

$$A = F_1 + F_2 + F_3 + \&c. + F_m$$
$$B = F_1 + 2 F_2 + 3 F_3 + \&c. + m F_m,$$

we have the equations

$$A - m x - \frac{m(m+1)}{2} N y = 0,$$
$$B - \frac{m(m+1)}{2} x - \frac{m(m+1)(2m+1)}{6} N y = 0.$$

Solving these, we find

$$x = \frac{2 + 4m}{m^2 - m} A - \frac{6}{m^2 - m} B.$$

$$y = \frac{12}{m^3 - m} \frac{B}{N} - \frac{6}{m^2 - m} \frac{A}{N}$$

In the present case,'

$m = 8 \qquad N = 14, \qquad A = 138\cdot 4, \qquad B = 770\cdot 9;$
whence $x = 1\cdot 44$
$y = 0\cdot 252,$

and we have the formula,

$$F = 1\cdot 44 + 0\cdot 252\ R.$$

APPENDIX II.

DETAILS OF THE WILLIS APPARATUS USED IN ILLUSTRATING THE FOREGOING LECTURES.

THE ultimate parts of the various contrivances figured in this volume are mainly those invented by the late Professor Willis of Cambridge. They are minutely described and illustrated in a work written by him for the purpose under the title *System of Apparatus for the use of Lecturers and Experimenters in Mechanical Philosophy*, London, Weale & Co., 1851. This work has long been out of print. It may therefore be convenient if I give here a brief account of those parts of this admirable apparatus that I have found especially useful. The illustrations have been copied from the plates in Professor Willis' book.[1]

The Willis system provides the means for putting versatile framework together with or without revolving gear for the purpose of mechanical illustration. Many parts which enter into the construction of the machine used at the lecture to-day will reappear to-morrow as essential parts of some totally different contrivance. The parts are sufficiently substantial to work thoroughly well. The scantlings and dimensions generally have

[1] I ought to acknowledge the kindness with which Mr. J. Willis Clark, of Cambridge, the literary executor of Professor Willis, has responded to my queries, while I am also under obligations to the courtesy of Messrs. Crosby, Lockwood, & Co.

been so chosen as to produce models readily visible to a large class.

It will of course be understood that every model contains some one or more *special* parts such as the punch and die in Fig. 73, or the spring balance in Fig. 17, or the pulley block in Fig. 33. But for the due exhibition of the operation of the machine a further quantity of ordinary framework and of moving mechanism is usually necessary. This material, which may be regarded as of a *general* type, it is the function of the Willis system to provide.

THE BOLTS.—The system mainly owes its versatility and its steadiness to the use of the iron screw bolt for all attachments. The bolts used are $\frac{3}{8}$" diameter; the shape of the head is hemispherical and the shank must be square for a short distance from the head so that the bolt cannot turn round when passed through the slits of the *brackets* or *rectangles*. When the head of the bolt bears on a slit in one of the wooden pieces a circular iron washer 2" in diameter, or a square washer 2" on each side, is necessary to protect the wood from crushing. There is to be a square hole in the washer to receive the square shank of the bolt and the thickness of the washers should be $\frac{1}{8}$". The nut is square or hexagonal, and should *always* have a washer underneath when screwed home with a spanner or screw-wrench. The most useful lengths are 2", 4", 6". The proper kind are known commercially as *coach-bolts*, and they should be chosen with easy screws, for facility in erecting or modifying apparatus. At least two dozen of the intermediate size and a dozen of each of the others are required. For elaborate contrivances many more will be necessary.

THE BEDS.—The simplest as well as the longest parts of the framework are called "beds" (Fig. 104). Each bed is made of two wooden bars. These bars are united by strong screws passing through small blocks of hard wood so as to keep the bars full $\frac{3}{4}$" asunder, and thus allow the shanks of the bolts to pass freely through the slit. The scantling of each bar is $2\frac{1}{2}$" × $1\frac{1}{2}$", and the beds are of various lengths from 1' to 10'

APPENDIX. 347

or even longer. The beds can be attached together in any required position by bolts 6" long. The rectangles and the brackets are attached to the beds by 4" bolts. In one con-

THE BED.

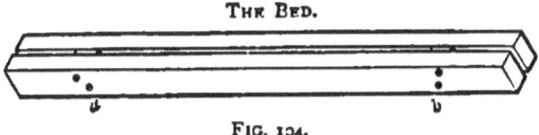

FIG. 104.

junction or another the beds will be found represented in almost every figure in the book. We may specially refer to Figs. 20, 44, 48, 49, 50, 65, 83.

THE STOOL.—Most of the larger pieces of apparatus have the *stool* as their foundation (see Figs. 11, 39, 102). It is often

THE STOOL.

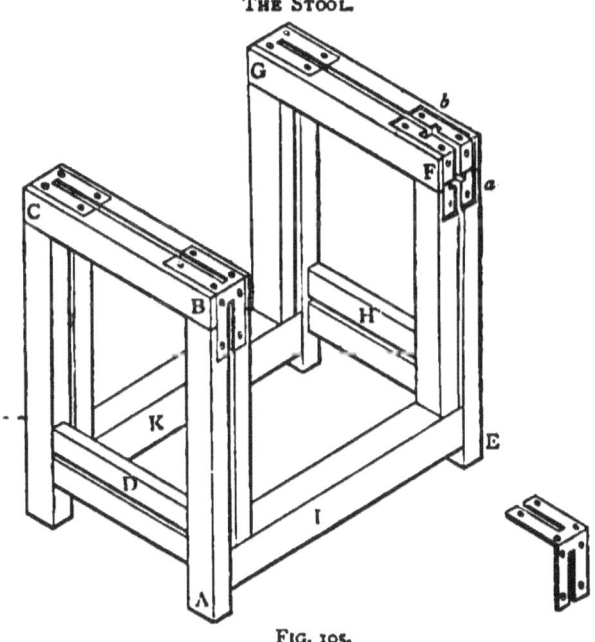

FIG. 105.

convenient as in Fig. 65 to employ a pair of stools, while one stool superposed on another gives the convenient stand in

Fig. 80. The stool is a stout wooden frame, providing a choice of slits to which beds or other pieces may be attached by bolts. The structure of the frame is shown in Fig. 105. It is 2′ 6″ high and its extreme horizontal dimensions are 2′ 6″ × 1′ 9″ of which the greater is A E. In other words, the longer sides of the stool are those open at the top. Each top corner is strengthened by an iron plate of which a separate sketch is shown. The scantlings of the parts of the stool are as follows:—The legs and horizontal top rails are 3″ × 2⅛″. Two of these rails with the intervening ⅜″ slit make the top and legs to be 4⅝″ wide. The bottom front rail I is 3″ wide and 4″ deep. The double side rails D, H are 1¾″ wide and 2½″ deep, being made thinner than the legs into which they are mortised in order to allow the washers of the bolts to pass behind them. The slits are to be full ⅜″ wide throughout. Beech or birch are very suitable materials, but softer woods will answer if large washers are invariably used.

THE RECTANGLE.—The useful element of the Willis system known by this name is of iron cast in one piece (Fig. 106). The rectangles are used in the attachment of beds to each other under special conditions, or they are often attached to the stools or to brackets. Indeed their uses are multifarious, see for examples Figs. 12, 58, 62, 89, 97, 102 and many others. The faces of the rectangle are 2½″ broad. The outside dimensions are 6″ and 9″, and the thickness of metal is ⅜″. Each side of the rectangle has the usual bolt slit ⅜″ clear. Rectangles of a larger size are often found useful, their weight makes them effective stands (see Figs. 35, 43, 52, 65).

THE RECTANGLE.

FIG. 106.

THE TOOTHED WHEEL.—The most convenient type of toothed wheel for our present purpose is that known as the cast-iron *ten-pitch*. In all such wheels the number of teeth is simply ten times the number of inches in the diameter. For example a wheel with 120 teeth is 12 inches in diameter. A number of ten-pitch wheels large and small must be pro-

APPENDIX. 349

vided. The actual assortment that will be necessary depends upon circumstances. For most purposes it will be sufficient to have the multiples of 5 from 25 upwards to 120, and then a few larger sizes such as 150, 180, 200. Duplicates of the constantly recurring numbers such as 30, 60, 120 are convenient. *Arm* wheels are always preferable to *plate* wheels in lightness and appearance as well as in price. All wheels are to be 1" thick at the boss which is faced in the latter at each side, and bored with a hole full 1" diameter, in which a key groove is cut. A pair of mitre wheels such as are used in Fig. 80 are sometimes useful.

THE PULLEY.—We have frequent occasion to use the pulley for conveying a cord, and a somewhat varied stock is convenient. Thus light brass pulleys are used in the apparatus shown in Fig. 3, and a stout pulley in Fig. 71. A cast-iron pulley about 10" in diameter is seen in Figs. 32 and 34. It is bored 1" in diameter with a key groove, and the boss is 1" thick. Some small pulley blocks similar to those used on yachts are often very useful.

THE STUD-SOCKET.—For mounting toothed wheels on the larger pulleys or for almost any rotating or oscillating pieces the stud-socket is used (see Fig. 107). The socket A B may be made of brass or of cast-iron. It is 1" in diameter so as to pass through the bosses of the wheels that have been bored to 1" with this object :—The socket is provided with a shoulder at one end (A) which is 1½" diameter, and with a strong screw B and octagonal nut at the other end. The extreme length of the socket is 3½", and the plain part of the 1" cylinder is 1¾" long. When two wheels are placed on the socket each of which has a boss 1" thick, the tightening of the nut will secure the wheels against the shoulder. A feather is screwed on the plain part which enters the key grooves in the wheels, and thus ensures that the wheels shall turn together. This feather should

THE STUD-SOCKET.

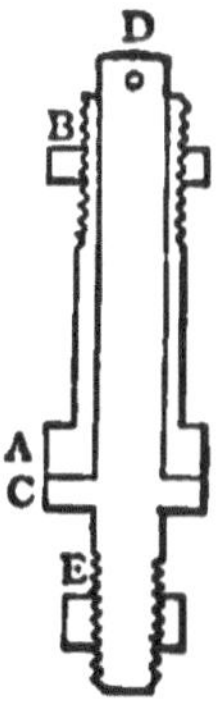

FIG. 107.

be small enough to slip *easily* into the key groove. If only a single wheel or if any peculiar piece such as a wooden cam or a disk of sheet iron has to be mounted, then collars or large thick washers must be placed on the socket so as permit the screw to bind the whole together. The socket revolves upon a stout iron stud C D, which is $\frac{5}{8}''$ in diameter. It bears a shoulder or flange C at the back of the same diameter as the base of the socket. The stud bears on the other side of the shoulder a strong screw and nut which project $1\frac{5}{8}''$ so as to allow the stud to be secured in a hole $1''$ deep in one of the brackets (to be presently described). The plain part of this screw near the shoulder must be $\frac{5}{8}''$ diameter. The front end of the stud is pierced with a hole to receive a spring pin to keep the socket from sliding off the stud. Among the many applications of the stud socket we may mention those shown in Figs. 30, 73, 74.

THE BRACKET.—There are six different forms of cast-iron brackets represented in the adjoining figures (Figs. 108—113).

The brackets are primarily intended as the supports of the stud-sockets. For this purpose each has a head $1''$ thick bored

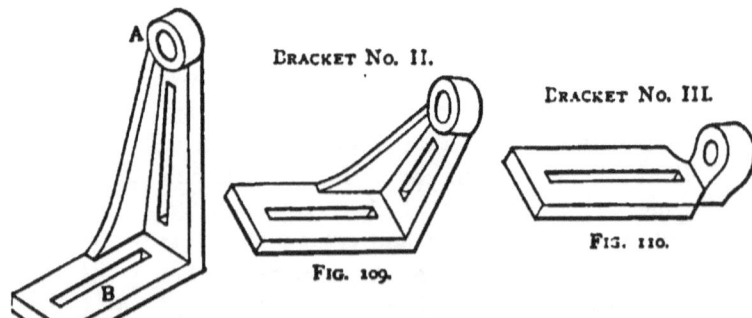

BRACKET No. I.
FIG. 108.

BRACKET No. II.
FIG. 109.

BRACKET No. III.
FIG. 110.

with a hole $\frac{5}{8}''$ diameter, and thus fitted to receive the screw on any of the studs. Each bracket stands on a base or *sole* with a slit full $\frac{3}{8}''$ wide for the bolts. The thickness of the sole is $\frac{5}{8}''$. The larger of the brackets I., II., and IV. have also slits in their

APPENDIX. 351

vertical faces. Brackets can be fastened either to the stool or to the beds or rectangles, and the variety of their forms enables the wheel-work carried on the stud sockets to be disposed in any

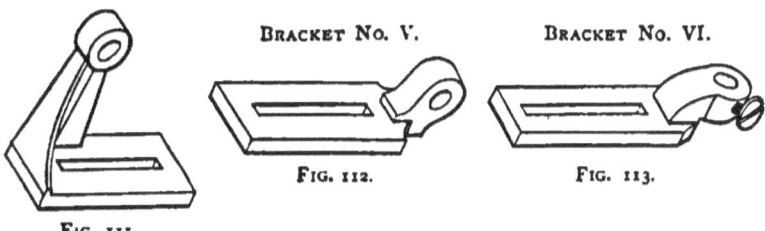

BRACKET No. IV.
FIG. 111.

BRACKET No. V.
FIG. 112.

BRACKET No. VI.
FIG. 113.

desired fashion. Brackets avail for many other purposes besides those of supporting rotating mechanism. (Look at Figs. 11, 12, 17, 20, 33, 38, 39, 73 and many others.)

THE SHAFTS AND TUBE-FITTINGS.—The stud sockets will not provide for every case in which wheels have to be mounted and driven. We must often employ shafts (see for instance Figs. 30, 47, 101). The shafts we use are turned iron rods $\frac{3}{4}$" in diameter, and of various lengths from 6" up to 4'. To support the shafts we use for bearings the *tube fitting* (Fig. 114). This is a brass casting which consists of a tube M N 2" long, and 1¼" in external diameter, bored $\frac{3}{4}$" so as to fit the shaft. The back of this tube is a flat surface parallel to the bore, and from it projects a screw $\frac{3}{8}$" diameter, and 1$\frac{3}{8}$" long with a nut which is however omitted in the drawing.

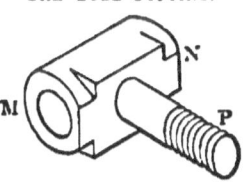

THE TUBE FITTING.
FIG. 114.

This screw may be of the same size as that of the studs, and it is intended for the same purpose, namely to attach the bearing to the hole in a bracket. The tube may of course be fixed at any desired angle in the plane parallel to the face of the bracket. To prevent the endlong motion of the shaft cast-iron or brass rings are employed (Fig. 115). These are bored $\frac{3}{4}$", and furnished with a binding screw by which they may be tightened on the shaft in any position. To avoid injury to the shaft it is well to have a narrow flat surface filed along it to receive the end of the binding screw. The use of the rings is shown in Fig. 47. If as often happens (see for example Fig. 102) a barrel has to be set in motion by a shaft the required attachment can be made by simply slipping on the barrel, and then putting at each end of it two of the pinned rings (Fig. 115). The pins enter holes bored into the barrel for their reception so that when the rings are bound to the shaft by their screws the barrel must revolve with the shaft.

THE PINNED RING.

FIG. 115.

THE ADAPTER.—For the attachment of wheels or other rotating pieces to the shaft an adapter is employed (Fig. 116). It is bored with a $\frac{3}{4}$" hole to fit the shaft, and the external diameter is 1". At one end is a shoulder through which the binding screw is tapped, and there is a nut and screw at the opposite end. A feather will prevent the wheel from turning round on the adapter which is itself made to revolve with the shaft by screwing the binding screw down on the shaft. Some adapters are only large enough for a single wheel 1" thick in the boss, but it is useful to have others that will take two wheels. Adapters are shown in use in Figs. 46 and 101.

THE ADAPTER.

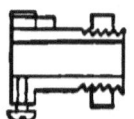

FIG. 116.

THE LEVER ARM. – To give motion to the mechanism a lever arm with a handle is frequently required (Fig. 117). It is bored 1" and has a key groove, and the hole is 1" long, so that the lever arm can be fixed on a stud socket like a wheel. By

THE LEVER ARM.

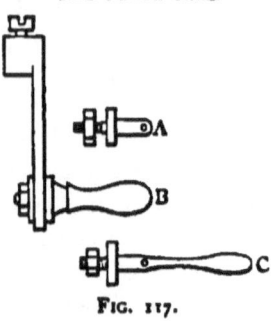

FIG. 117.

APPENDIX. 353

the aid of an adapter the lever arm is attached to a shaft. For the use of the handle see Figs. 30 and 101. There are however many other uses to which the lever arm is occasionally put. It can be used as a crank, and in linkage arrangements a pair of lever arms are very convenient. Studs A or C can replace the handle when necessary.

Such are the parts of the Willis apparatus which are adapted for our present purpose. It remains to add that the fits should be very easy, and the parts should be readily interchangeable.

INDEX.

A.

Accident, risk of, 32
Action, 6
Adapter, Willis apparatus, 352
Angle of friction, 78
  of statical friction, 80
Apparatus for centre of gravity, 62
  for equilibrium of three forces, 7
  to show friction, 65, 78
  the Willis, 345
Appendix I., 339
Attwood's machine, 232
Axes, permanent, 279

B.

Balance, defective, 48
  spring, 16
Bar, equilibrium of a, 38
Bat, cricket, 309
Beam, breadth of, 193
  breaking load of, 193, 196
  cast iron, 222
  collapse of, 186
  deflection of, 179
  elasticity of, 184
  load on, 197
  placed edgewise, 193
  strained, 178
  strength of, 190
  uniformly loaded, 198
  with both ends secured, 200
  with one end secured, 201
Beds in Willis apparatus, 346
Bob, raising or lowering the, 320
Bolts, use of, in Willis apparatus, 346
Bracket, Willis apparatus, 350
Brass, specific gravity of, 56
Breaking load, 177

Bridge, deflection of, 208
  mechanics of, 218
  Menai, 218
  suspension, 225
  the Wye, 215
  tubular, 223
  with four struts, 210
  two struts, 206
  two ties, 211
Brunel, Sir J., the Wye bridge, 215

C.

Capstan, 151
Cast-iron beam, 222
Catenary, 226
Cathetometer, 180
Centre of gravity, 57
  of a wheel, 61
  position of, 59
  oscillation, 304
  percussion, 307
Circular motion, 267
  action of, 271
  applications of, 276
  cause of, 270
  in governor balls, 276
  in sugar refining, 276
  nature of, 267
  on liquids, 271
  on the earth, 276
Circular pendulum, 284
Clamps, 203
Clock pendulum, 299
  principles of, 318
  rate of, 322
Coefficient of friction, 74, 82
Collapse of a beam, 186
Compensating pendulum, 319

A A 2

INDEX.

Composition of forces, 1, 9
  parallel forces, 35, 37, 42
  vibrations, 299, 315
Conical pendulum, 310
Couple, 44
Crane, 29, 162
  friction in, 166
  mechanical efficiency of, 165
    Table XXI. 165
      XXII. 166
  velocity, ratio of, 163
Cricket bat, 309
Crowbar, 123
Cycloid, 295

D.

Dead-beat escapement, 328
Definition of force, 2
Deflection of a beam, Table XXIII. 182
Differential pulley, 112
  Table XI. 114
Direction of a force, 5

E.

Eade, Mr., epicycloidal pulley block, 116
Easter Island, 100
Elasticity of a beam, 184
Energy, 85, 94
  storage of, 256, 258
  unit of, 95
Engine, locomotive, 83
Epicycloidal pulley block, 80, 116
  Table XII. 118
Equilibrium, neutral, 61
  of a bar, 38, 41
    three forces, 6
    two forces, 6
  stable, 59
  unstable, 59
Escapement, 324
  dead-beat, 328
  recoil, 328
Expansion of bodies, 321
Experiment by M. Plateau, 273

F.

Fall in a second, 239
Falling body, motion of, 230
Feet, how represented, 7
Fibres in state of compression, 184
  tension, 184
First law of motion, 230

Flywheel, 260
  in steam-engine, 262
Foot pound, 95
Force, a small, and two larger, 12
  definition of, 2
  destroying motion, 3
  direction of a, 5
  magnitude of a, 4
  measurement of, 4
  of friction, 65
    gravity, 50
  one, resolved into three, 26
    two, 17
  representation of, 5
  standard of, 4
Forces, composition of, 1, 9
  equilibrium of three, 6
    two, 6
  illustrations of, 3
  in inclined plane, 136
  parallel, 34
  parallelogram of, 10
  resolution of, 16
Formula for pulley block, 109, 114
Framework, 203, 345
Friction, 65
  accurate law of, 75
  a force, 66
  and pressure, 72
  angle of, 78
  angle of statical, 80
  apparatus to show, 65, 68, 78
  caused by roughness, 66
  coefficient of, 74, 82
  diminished, 66
  excessive, 115
  experimenting on, 66
  in crane, 166
    differeutial pulley block, 113
    inclined plane, 132
    lever, 123
    pulleys, 89
      law of, 91
    rope and iron bar, 87
    wheel and axle, 153
    wheel and barrel, 158
  laws of, 73, 81, 82
  mean, 75
  motion impeded by, 70
  nature of, 65
  overcoming, 93
  Table I. 69
    II. 71
    III. 74
    IV. 76
    V. 78
    VI. 81
    VII. 81
    VIII. 81
  upon axle, 155
  wheels, 93

INDEX.

G.

Galileo and falling bodies, 235
  kinetics, 230
  the pendulum, 284
  tower of Pisa, 233
Gathering pallet, 336
Girder, 219
  as slight as possible, 221
Governor balls, 276
Graham, dead-beat escapement, 328
Graphical construction, 339
Gravity, 50
  action of, 243
  and the pendulum, 292
  and weight, 52
  centre of, 57
  defined, 246
  different effects of, 53
  independent of motion, 241
  in London, 292
  specific, 53
Grindstone, treadle of, 128

H.

Hammer, 252
  theory of the, 252
Hands of a clock, 331
Horse-power, 96

I.

Illustration of parallelogram of forces, 10
Illustrations of forces, 10
  resolution, 19
Inches, how represented, 7
Inclination of thread, 140
Inclined plane, 131
  forces on, 136
  friction in, 132
  mechanical efficiency of, 139
  Table XIII. 134
    XIV. 137
    XV. 138
  velocity, ratio of, 139
Inertia, 250
  inherent in matter, 252
Iron girders, 219
  specific gravity of, 55
Isochronous simple pendulum, 303
Ivory, specific gravity of, 56

J.

Jib, 29, 163

K.

Kater, Captain, 305
Kinetics, 230

L.

Large wheels, advantages of, 93
Law of falling bodies, 238
  friction in pulleys, 91
  lever of first order, 122
  pressure, 37
Laws of friction, 73, 81, 82
Lead, specific gravity of, 56
Leaning tower of Pisa, 233
Level, 56
Lever, 119
  and friction, 123
  applications of, 123
  arm, Willis apparatus, 352
  laws of, 130
  of first order, 119
    law of, 122
  of second order, 124
  of third order, 128
  weight of, 121
Lifting crane, 29
Line and plummet, 56
Load, breaking, 177
Locomotive engine, 83

M.

Machine, Attwood's, 232
  punching, 263
Machines, pile-driving, 255
Magnitude of a force, 4
Margin of safety, 33
Mass, 236
Mean frictions, 75
Measurement of force, 4
Mechanical powers, 85, 100
  apparatus, Willis, 345
Menai Bridge, 218
Method of least squares, 342
Moment, 130
Monkey, 257
Motion, first law of, 230
  of falling body, 230

N.

Neutral equilibrium, 61
Newton and gravity, 289
Nut, 140

O.

Oscillation, centre of, 304

P.

Pair of scales, 48
  testing, 48
Parabola, 226

358   INDEX.

Parallel forces, 34
　　composition of, 35, 37, 42
　　opposite, 44
　　resultant of, 43
Parallelogram of forces, 10
Path of a projectile, 247
Pendulum and gravity, 292
　　circular, 284
　　compensating, 319
　　compound, 299, 301
　　conical, 310
　　formula for, 292
　　Galileo and the, 286
　　isochronous simple, 303
　　length of the seconds, 292, 318
　　motion of the, 285
　　of a clock, 299
　　simple, 284
　　time of oscillation, 286, 289
Percussion, centre of, 307, 309
Permanent axes, 279
Pile-driving machines, 255
Plateau, M., experiment by, 273
Plummet, 56
Powers, mechanical, 85
Pressure and friction, 72
　　law of, 37
　　of a loaded beam, 35, 37
Principles of framework, 203
Projectile, path of, 247
Pulley block, 99
　　differential, 110
　　epicycloidal, 80
　　three sheave, 106
　　velocity, ratio of, 112
Pulley, ordinary form of, 86
　　single movable, 101
　　　　fixed, 86
　　use of, 88
　　velocity, ratio of, 103
Pulleys, friction in, 89
　　in windows, 86
　　in Willis apparatus, 349
Punching-machine, 263
　　force of, 265

R.

Rack, 334
Reaction, 6
Recoil escapement, 328
Rectangle in Willis apparatus, 348
Representation of a force, 5
Resistance to compression, 172, 175
　　extension, 172
Resolution of forces, 16
　　one force into three, 26
　　　　two, 17
Resultant, 9
　　of parallel forces, 43

Rings in Willis apparatus, 352
Risk of accident, 32

S.

Safety, margin of, 33
Sailing, 21
　　against the wind, 24
Scales, 46
Screw, 139
　　and wheel and axle, 167
　　form of, 139
　　Table XVI. 142
　　velocity, ratio of, 143
Screw-bolt and nut, 148
　　jack, 131, 145
　　　　Table XVII. 146
Second, fall in a, 239
Seconds, pendulum, 318
Shafts, Willis apparatus, 351
Shears, 126
Simple pendulum, 284
Single movable pulley, Table IX. 104
Snail, 334
Specific gravity, 53
　　of brass, 56
　　　iron, 55
　　　ivory, 56
　　　lead, 56
Spirit level, 56
Spring balance, 16
Stable equilibrium, 59, 282
Standard of force, 4
Statical friction, angle of, 80
Stool in Willis apparatus, 347
Storage of energy, 256, 258
Stored-up energy exhibited, 261
Strength of a beam, 190
Striking parts, 333
Structures, 169
Strut, 28
Stud socket in Willis apparatus, 349
Sugar refining, 276
Suspension bridge, 225
　　mechanics of, 225
　　tension in, 228

Table I. 69
　　II. 71
　　III. 74
　　IV. 76
　　V. 78
　　VI. 78
　　VII. 81
　　VIII. 81
　　IX. 104
　　X. 108

Table XI. 114
XII. 118
XIII. 134
XIV. 137
XV. 138
XVI. 142
XVII. 146
XVIII. 154
XIX. 159
XX. 162
XXI. 165
XXII. 166
XXIII. 182
XXIV. 190
Tacking, 25
Tension along a cord, 17
Three sheave pulley block, 106
Tie, 28, 175
    rod, 29, 32
Timber, bending, 171
    compression of, 172
    extension of, 172
    properties of, 170
    rings in, 171
    seasoning, 171
    warping, 171
Tin, 223
Toothed wheels, 160
Tower of Pisa, 233
Train of wheels, 330
Transverse strain, 181
Treadle of a grindstone, 128
Tripod, 28
    strength of, 28
Truss, simple form of, 212
Tube fittings, Willis apparatus, 351
Tubular bridge, 223

U.

Unstable equilibrium, 59, 282

V.

Velocity, 231
    ratio of inclined plane, 139
    pulley, 103
    pulley block, 112
    screw, 143
    wheel and axle, 152
    wheel and pinion, 161
Vibrations, composition of, 299, 315

W.

Wedge, 139
Weighing machines, 123
    scales, 46, 48
Weight caused by gravity, 52
    of water, 54
Wheel and axle, 149
    and differential pulley, 167
    screw, 167
    experiments on, 152
    formula for, 154
    friction in, 153
    Table XVIII. 154
    velocity, ratio of, 152
Wheel and barrel, 158
    formula for, 160
    friction in, 158
    Table XIX. 159
Wheel and pinion, 160
    efficiency of, 161
    Table XX. 162
    velocity, ratio of, 161
Wheel, centre of gravity of, 61
Wheels, 92
    friction, 93
Wheels in Willis apparatus, 348
Willis system of apparatus, 345
Winch, 151
Wind, direction of, 22
Work, 85, 94
Wye Bridge, 215

THE END.

RICHARD CLAY AND SONS, LIMITED, LONDON AND BUNGAY.

www.ingramcontent.com/pod-product-compliance
Lightning Source LLC
Chambersburg PA
CBHW031419230426
43668CB00007B/363